WILDFLOWERS OF NORTH AMERICA

A Coast-to-Coast Guide to More Than 500 Flowering Plants

DAMIAN FAGAN

GUILFORD, CONNECTICUT

FALCONGUIDES®

An imprint of Globe Pequot, the trade division of The Rowman & Littlefield Publishing Group, Inc.
4501 Forbes Blvd., Ste. 200
Lanham, MD 20706
www.rowman.com
Falcon and FalconGuides are registered trademarks and Make Adventure Your Story is a trademark of The Rowman & Littlefield Publishing Group, Inc.

Distributed by NATIONAL BOOK NETWORK

British Library Cataloguing in Publication Information available

Library of Congress Cataloging-in-Publication Data
Names: Fagan, Damian, author.
Title: Wildflowers of North America : a coast-to-coast guide to more than 600 flowering plants / Damian Fagan.
Other titles: Coast-to-coast guide to more than 600 flowering plants
Description: Guilford, Connecticut : FalconGuides, [2022] | Includes bibliographical references and index. | Summary: "A field guide to more than 500 wildflowers found in North America, organized by color and alphabetically to enable wildflower enthusiasts and nature lovers to identify and learn about the natural and cultural history of flowering plants"— Provided by publisher.
Identifiers: LCCN 2021035799 (print) | LCCN 2021035800 (ebook) | ISBN 9781493057818 (paperback) | ISBN 9781493057825 (epub)
Subjects: LCSH: Wild flowers—North America. | Flowers—North America—Identification. | Field guides.
Classification: LCC QK85.5 .F34 2022 (print) | LCC QK85.5 (ebook) | DDC 582.13097—dc23
LC record available at https://lccn.loc.gov/2021035799
LC ebook record available at https://lccn.loc.gov/2021035800

∞™ The paper used in this publication meets the minimum requirements of American National Standard for Information Sciences—Permanence of Paper for Printed Library Materials, ANSI/NISO Z39.48-1992.

THIS BOOK IS DEDICATED TO ALL THE WILDFLOWER ENTHUSIASTS OUT THERE: TAXONOMISTS, GARDENERS, PHOTOGRAPHERS, EDUCATORS, WILDCRAFTERS, EXPLORERS, NEWBIES, AND SEASONED PROS, FROM PAST TO PRESENT, WHO FIND ENJOYMENT AND WONDER IN THE WORLD OF WILDFLOWERS.

CONTENTS

ACKNOWLEDGMENTS

While working on this field guide, I truly felt like I was standing on the shoulders of giants. Many have contributed to the knowledge and understanding of plants, including the many Indigenous peoples who roamed North America long before Europeans ventured across the ocean. Their shared wisdom of medicinal, edible, and toxic plants provided many New World settlers with an introduction to this valuable information. Plant collectors soon followed, and the thirst for more knowledge unfolded like a spring blossom. Modern-day taxonomists use genetic sequencing to determine relationships of plants, and they occasionally shake the taxonomic tree and rearrange some of its branches and leaves representing the hierarchy of plant taxa. Ever evolving, this knowledge base is rooted in a relationship that spans the millennia and draws in humans and wildlife, particularly the pollinators, whose floral connections are millions of years in the making. So much to learn, so little time, but enough of a reason to stand still in a field of flowers and soak up the paradise that flits and flutters, blossoms and blooms all around us.

I am grateful to the FalconGuides staff for their willingness to let me work on this project and for all the help they have provided in terms of editing, layout, printing, distribution, and marketing. In particular, Katie O'Dell (acquiring editor), Meredith Dias (senior production editor), and Joanna Beyer (layout artist) have been very understanding and helpful in the production of this book.

I also appreciate all the photographers who provided images for use in this book or who share their work through iNaturalist and allow their images to be used royalty free. Photographer credit appears along with their image. This work is a shared effort.

INTRODUCTION

The third-largest continent on Earth, North America stretches more than 5,000 miles north to south from the Arctic Ocean to the Isthmus of Panama. The landform covers more than 9.5 million square miles and is bordered mostly by water—the Pacific Ocean to the west, the Atlantic Ocean and Caribbean Sea to the east, and the Arctic Ocean to the north—and by South America to the south. The continent includes Greenland and islands and territories located in the North Atlantic and Caribbean Sea as well.

This field guide focuses on wildflowers growing in the contiguous United States and Canadian provinces. Some plants in the guide extend into Mexico, Latin America, or Alaska, and these distributions are noted in the plant descriptions. This guide focuses on herbaceous, nonwoody plants, excluding shrubs and trees. With more than 17,000 vascular plants growing in North America, not including invasive species, this field guide covers representatives of different regions and plant families.

Biogeography of North America

Five broad physical regions divide the North American continent: the western mountains, the Great Plains, the Canadian Shield, the diverse eastern region, and the Caribbean. Parts of Mexico are considered an extension of the western United States and are included in the definition of North America. However, due to the scale of this book, plants found growing only south of the US-Mexico border have not been included.

Across these regions and elevational extremes is a diverse and intriguing landscape that includes mountain ranges, arid deserts, open tundra, verdant forests, endless prairies, dissecting rivers, lowland swamps, and coastal reaches. One simple version of classifying these habitats is to lump them into six categories, called biomes: forest, grassland, freshwater, marine, desert, and tundra. This general classification ignores a more refined approach based on specific habitat types; here are a few examples: coniferous forest, sagebrush-shrublands, ponderosa pine woodlands, alkali sinks, freshwater marshes. This guide utilizes a mixture of both generic and specific biomes and habitat types to describe areas where a specific wildflower grows.

Topography and Climate

In addition to a wide range of habitats found throughout North America, variations in topography, soils, and climate also affect plant distribution and species diversity. Elevational extremes range from 20,310 feet on the summit of Denali in Alaska to 283 feet below sea level in the desert basins of California's Death Valley. Between these extremes lies a tapestry of topographical features bisected by mountain ranges and river basins. Diverse geologic features include sedimentary, igneous, and metamorphic rocks. Erosion of these layers creates the soil foundation upon which plants grow. In some areas, the resultant soil type limits plants that can grow there, resulting in a unique assemblage of plants with a limited range known as endemics.

Temperature and climatic conditions also affect plant distribution. Temperature extremes range from 134°F to -80°F, and average annual rainfall varies from less than 5 inches in the American Southwest's Mohave Desert to more than 170 inches in Washington State's Hoh Rain Forest. The ability of plants to survive under these conditions is what makes them so interesting—both in form and in function.

This guide contains a mixture of wildflower species with broad ranges along with some more-limited in their distribution. Additional closely related species of interest are included in the "Comments" section.

How to Use this Guide

This guide is primarily organized by flower color. Though arranged by color, such as Blue, Green, Red, White, and Yellow, there are noted variations within each broad division that will help identify the species. For example, the "Blue Flowers" section includes lavender and purple flowers.

Within each section, the following hierarchy has been employed to create a sense of order: Family, Genus, Species, Common Name. For wildflower enthusiasts new to plant identification, knowing something of a plant family's characteristics helps create a starting point when identifying an unknown plant. For example, members of the Sunflower family (Asteraceae) have flowers composed of ray and/or disk flowers.

The Linnaean System

Swedish naturalist Carl von Linné, aka Linnaeus (1707–1778), derived the current binomial system by which plants, and all other living creatures, are taxonomically arranged by genus and species. Genera with similar floral features are then arranged into families, a higher level of organization. This Linnaean System created order out of chaos, for during his time the scientific naming of plants and animals had few rules and little standardization in organizing this information. His system provided a common science-based language of latinized names that would be consistent for scientifically naming plants and animals. For example, butterfly milkweed (*Asclepias tuberosa*) is in the Dogbane, or Milkweed, family Apocynaceae. *Asclepias* is the generic or genus name; *tuberosa* is the specific epithet or species name. Here is one format of the plant's taxonomic hierarchy:

Kingdom: Plantae—Plants
Subkingdom: Tracheobionta—Vascular plants
Super division: Spermatophyta—Seed plants
Division: Magnoliophyta—Flowering plants
Class: Magnoliopsida—Dicotyledons
Subclass: Asteridae
Order: Gentianales—Dogbane/Gentian/Milkweed
Family: Apocynaceae—Milkweed family
Genus: *Asclepias*—Milkweed
Species: *Asclepias tuberosa*—Butterfly milkweed

Note: Not all botanists agree on the taxonomic arrangement or naming for all plants. Modern-day taxonomists incorporate genetic research to reveal identification and relationships of plants to one another instead of relying solely on morphological characteristics of flowers, leaves, and fruits. Staying current with taxonomic changes is a challenge, and some plants may be better known through their synonyms.

This new research has resulted in taxonomic changes at the species, genera, and family levels. Synonyms—alternative scientific names that a particular species used to be identified by—are sometimes included to facilitate these nomenclatural differences. An example of this is scarlet gilia (*Ipomopsis aggregata*), which used to be named *Gilia aggregata.* As DNA sequencing research continues, future changes to botanical nomenclature are certain to occur.

Common names often vary from region to region, and there is no standardization for these titles. Additional common names will occasionally be mentioned in the "Comments" section along with information, such as the etymology of the Latin or Greek scientific words, that reveals aspects of these wildflowers. Remember, too, that plants often exhibit a range of characteristics and that flowering seasonality may vary between regions and across elevations.

Conservation of Wildflowers

Illegal harvesting of certain species, especially in the Cactus and Orchid families, has resulted in a dramatic impact to those species. Other factors include the introduction of noxious weeds, habitat alteration, herbicide misuse, wildland fire, and other destructive actions. Some plants with wide distributions may weather these impacts, but others with more restrictive ranges (i.e., endemics or rare plants) may not. Please refrain from harming or destroying wildflowers, leaving them intact for others to enjoy.

Many organizations and individuals are involved in conserving wildflowers throughout North America and other parts of the globe. Supporting the endeavors of these entities is a worthwhile cause and a way to have a lasting impact on the legacy of wildflower enjoyment.

For gardeners creating a wildflower garden, look for seeds that were harvested locally to maintain genetic integrity, or select plants that were grown locally to avoid inadvertent introduction of diseases or insects to a new area. Many organizations can also help with selecting wildflowers that benefit pollinators in your neighborhood. Just think what a difference you can make by removing a 10-foot-square patch of lawn and replanting it with habitat for native pollinators such as bees, butterflies, and beetles!

PLANT CHARACTERISTICS

Most of the plants described in this field guide are commonly referred to as wildflowers, or forbs—herbaceous flowering plants that are not a grass, rush, or sedge. This category of plants encompasses annuals and perennials, plants with a life span that is either one year or greater than one year, respectively. A few species are biennials, plants that grow leaves the first year and produce flowers the second year. The decision of which wildflowers to include in this guide was based primarily on providing distribution and selecting representatives from across the country. An occasional endemic or rare plant is included, but since this guide is more an overview of wildflowers in North America, wildflower enthusiasts may need to check a local source for plants specific to a region and not included in this guide.

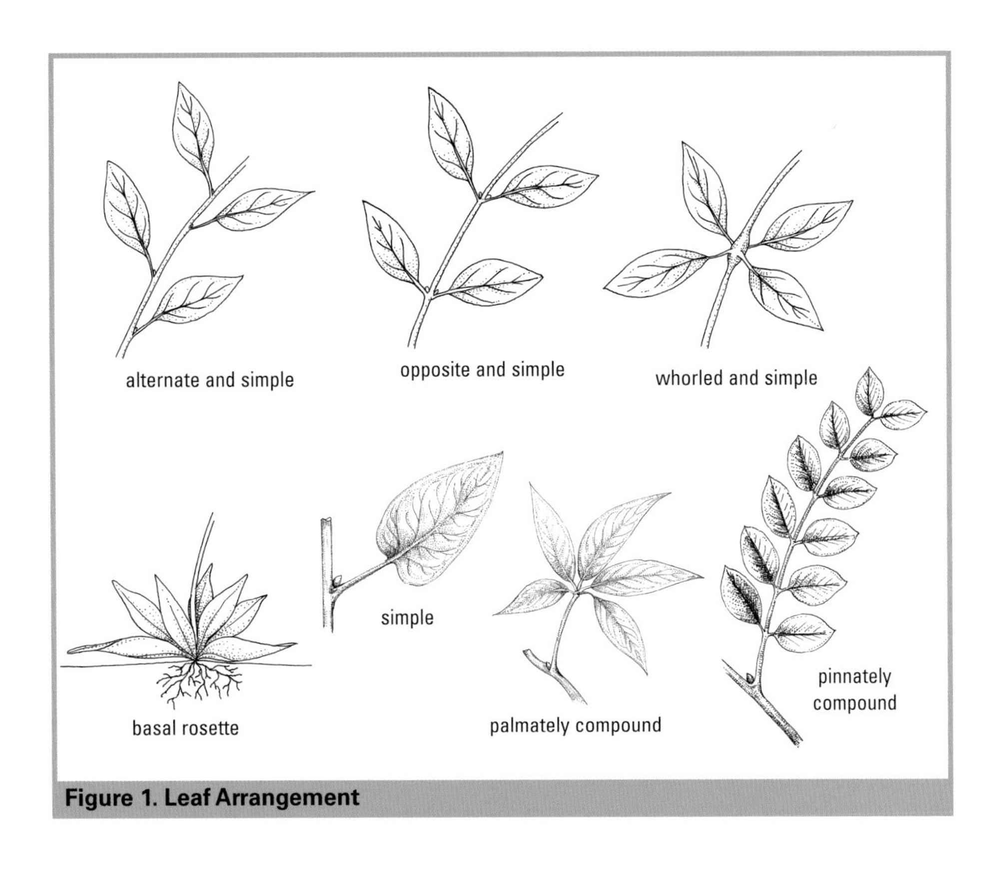

Figure 1. Leaf Arrangement

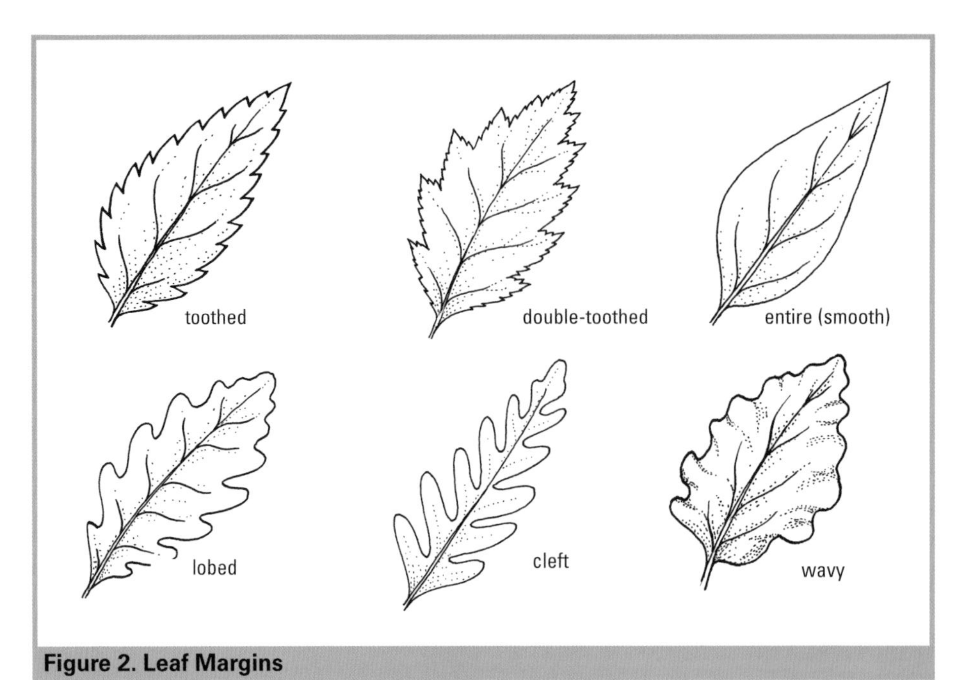

Figure 2. Leaf Margins

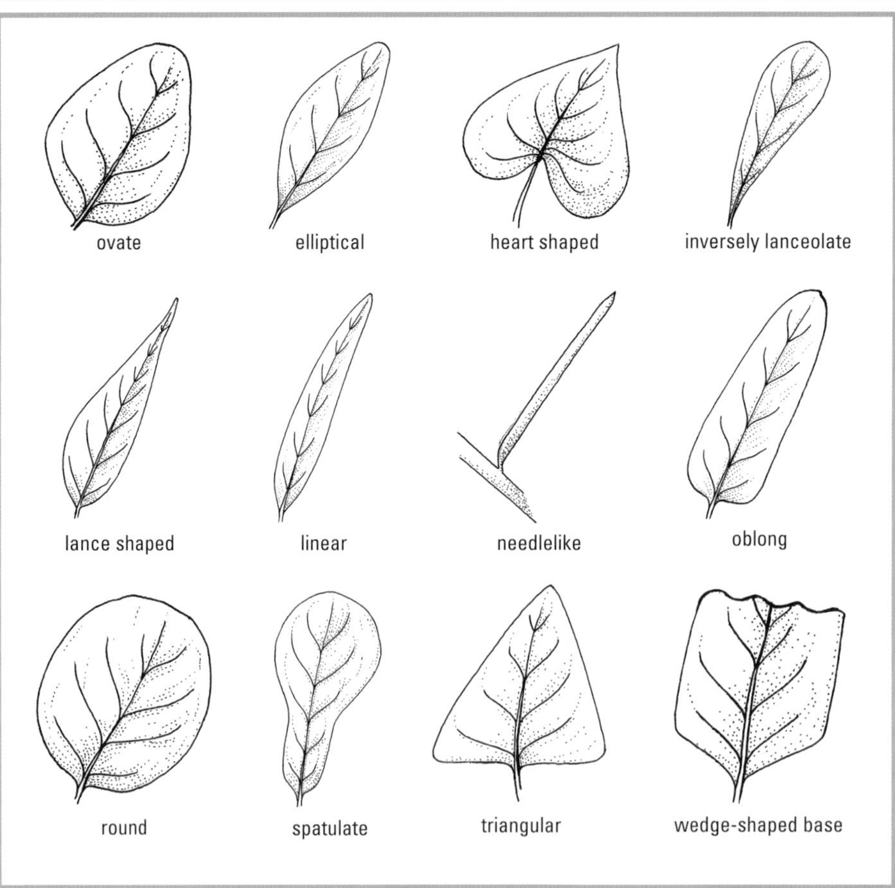

Figure 3. Leaf Shapes

Leaf Structure

The illustrations help define some of the terms used in this field guide that describe leaf and flower shapes or features. Though not an all-inclusive list, the illustrations help visually define some characteristics of leaf shape, arrangement, and flower structure. Technical terms have been kept to a minimum; for their definitions, see the "Glossary." Several good botanical manuals that describe leaf and flower characteristics are listed in the "Resources" section.

Observing the structure and arrangement of leaves may aid in plant identification. Key features to observe include the arrangement of the leaves off a main stem, their margins, and their overall shape. The following figures visually define these characteristics.

Flower Structure

Flowers represent the sexual form of the plants in this guide. Encased within protective petals or sepals are the male and female reproductive parts—stamens and pistils, respectively. Figure 4 shows the general structure of a flower; however, not all flowers are created equally. The variation in structure, floral parts, and pollinator lures has resulted in the unique assemblage of how flowers appear today. There is no normal; literally, variation is the spice of a wildflower's life.

In general, a flower's reproductive organs are enclosed within outer layers of sepals and/or petals. Sepals may be green and inconspicuous or, in the case of certain lilies, be similar in appearance to petals so that the parts are called tepals. The collective of sepals is called the calyx.

Within the calyx is another layer composed of petals. These often colorful and showy petals may be fused or separate; collectively they form the corolla. The corolla is the "flower," unique in shape and color and pattern, which functions to protect the inner sex organs and visually attract pollinators.

Inside the corolla of wildflowers are stamens, the pollen-producing structures of a flower, and the pistils, or seed-producing structures. This "one-house" structure is known as monoecious. (Some shrubs and trees produce separate male and female flowers, a dioecious structure.)

Often long and thin, the number of stamens present in a flower varies, and at their tip is a club-like or elongated appendage called the anther. Pollen is released from the anther and, for species such as penstemons, is an important characteristic to aid in species identification.

Three parts compose the pistil: the stigma, the style, and the ovary. The stigma is the pollen receptor that sits atop a stalklike structure called the style, which connects the stigma to the ovary housing the ovules—structures that will become the seeds, fruits, or berries after fertilization. Sometimes the structure of the stigma and style aids in identification; carrying a hand lens or magnifying glass may be needed to observe these structures.

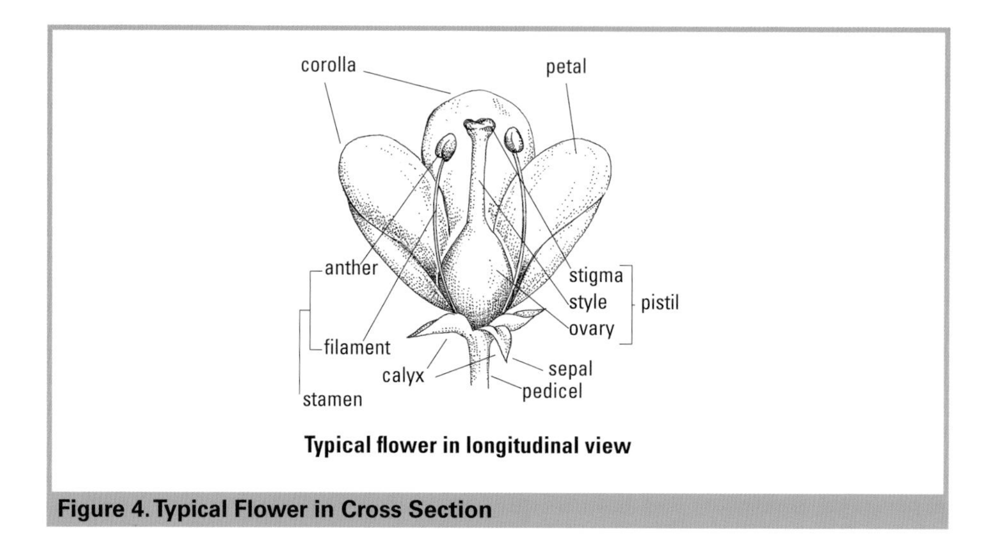

Figure 4. Typical Flower in Cross Section

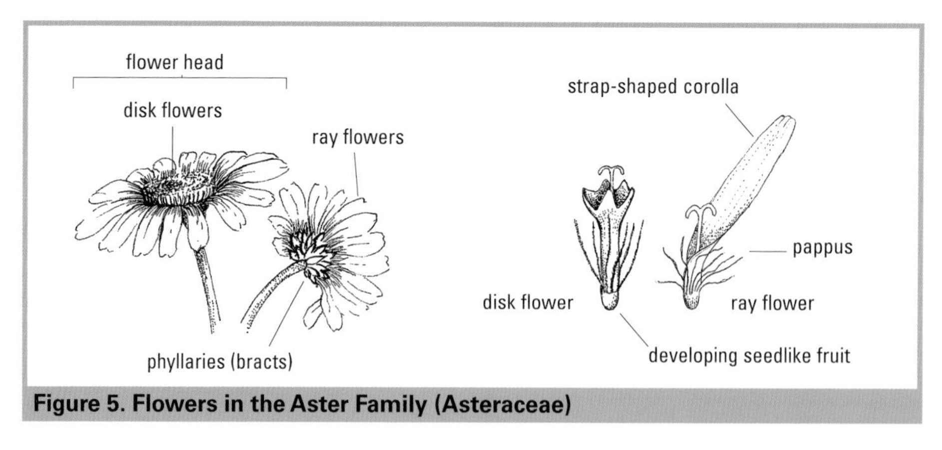

Figure 5. Flowers in the Aster Family (Asteraceae)

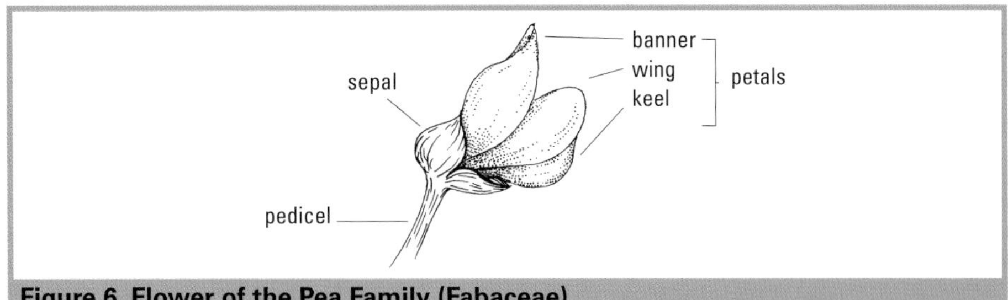

Figure 6. Flower of the Pea Family (Fabaceae)

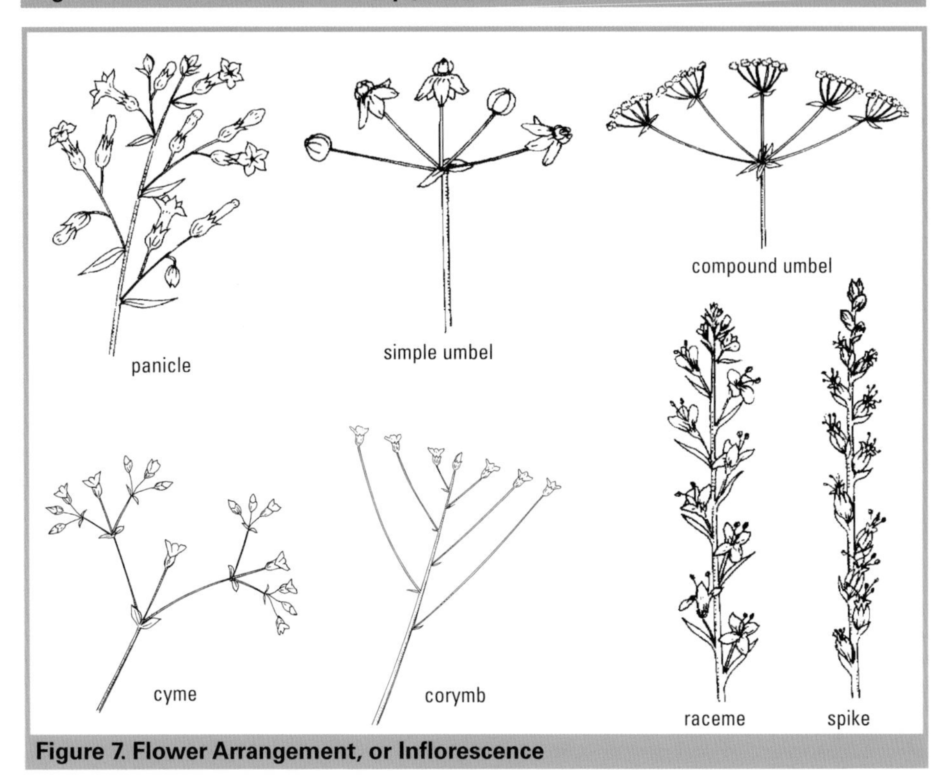

Figure 7. Flower Arrangement, or Inflorescence

BLUE FLOWERS

STEVE R. TURNER

This section includes plants with light to dark blue flowers and those ranging from lavender to purple. Occasionally, due to genetic variation, a blue-flowered plant, such as common camas, may have white flowers. Flowers that fall into the magenta, pink, or rose range may be in the "Red and Violet Flowers" section.

TOM LEBSACK

THICKLEAF AMERICAN WATERWILLOW

Justicia americana
Acanthus family (Acanthaceae)

Description: Perennial and colony-forming, grows 1'–3' tall, spreads by rhizomes. Stalkless leaves are linear or lance-shaped, opposite, 2"–6" long, mostly smooth along the margins but occasionally with rounded teeth on the margins. Two-lipped flowers are borne on long stalks, up to 4" long, in clusters. Flowers are about ¾" wide and bicolored; the upper lip is light violet to white and arches over a lower white petal with purple mottling; 2 lateral petals are spreading outward and have some purple spotting. Anthers are purplish red. Fruit is a brown capsule.

Bloom season: Late spring to early fall

Range/habitat: Eastern North America in partially submerged waters along rivers and lake shorelines

Comments: *Justicia* honors James Justice (1698–1763), a Scottish horticulturalist and gardener who published several guides to gardening in Scotland and England. *Americana* ("of America") refers to the distribution of this plant. Hummingbird bush (*J. californica*), also called chuparosa, bears scarlet flowers and grows in the Southwest. The roots and rhizomes provide aquatic habitat for fish and invertebrates.

ELEANOR DIETRICH

CAROLINA WILD PETUNIA

Ruellia caroliniensis
Acanthus family (Acanthaceae)

Description: Perennial, 2'–3' tall. Leaves are opposite, spoon-shaped to oval, crowded along the short stem. Blue-violet to white, funnel-shaped flowers with long tube and 5 spreading petals are borne in small clusters or solitary on short hairy stems; flowers have deep purple veins. Calyx has 5 long-pointed lobes. Seeds have hooked projections.

Bloom season: Late spring to fall

Range/habitat: Eastern and southern United States in lawns, moist woods, meadows, and roadsides

Comments: *Ruellia* honors Jean de la Ruelle (1479–1537), a French physician and botanist; *caroliniensis* ("of or from Carolinas") refers to the type locality of this plant. The flowers, which resemble but are unrelated to garden petunias, last for 1 day, but the blooming season is long. Butterflies are attracted as pollinators to these flowers and are rewarded with abundant nectar.

STEVE R. TURNER

EASTERN BLUESTAR

Amsonia tabernaemontana
Dogbane family (Apocynaceae)

Description: Perennial, 1'–3' tall. Leaves and stems ooze a milky latex when cut. Alternate, lance- to oval-shaped leaves are up to 6" long, generally hairless, and have smooth edges. Bluish-white tubular flowers with 5 spreading lobes arise in clusters along an elongated, leafy stalk. Flowers are ½"–¾" wide and 1" long. Narrow throat of the flower has white hairs that face inward. Fruit is a 4"–5" long, thin pod; usually 2 per flower.

Bloom season: Early spring to early summer

Range/habitat: Central United States in open woods, meadows, and stream banks

Comments: *Amsonia* honors Dr. John Amson (1698–1765?), an English physician and botanist who settled in Virginia and who treated George Washington in 1758 during the French and Indian War. *Tabernaemontana* honors Jacobus Theodor von Bergzabern (1525–1590) (Tabernaemontanus was his Latinized name), German botanist and physician, often called the "father of German botany." Ruby-throated hummingbirds and long-tongued butterflies are attracted to the flowers as pollinators. Woolly bluestar (*A. tomentosa*), a close relative, grows in the Southwest. Also known as blue dogbane.

ELEANOR DIETRICH

FLORIDA MILKVINE

Matelea floridana
Dogbane family (Apocynaceae)

Description: Deciduous twining vine up to 10' long. Heart-shaped leaves are up to 4" long, oppositely arranged, and exude a milky latex when cut. Flowers are ¾" long, with 5 narrow reddish-purple to blackish-purple petals arising from a central whitish ring. Fruit is a long, narrow spiny seedpod; seeds have silky hairs.

Bloom season: Late spring and summer

Range/habitat: Mainly Florida Panhandle and southern Georgia in sandhills, woodlands, and open habitat

Comments: *Matelea* may mean "swooning" or "fainting," but the derivation is unclear. *Floridana* ("of or from Florida") refers to the distribution of the plant. This Florida state endangered species is a larval host plant for monarch, queen, and soldier butterflies, and the flowers are primarily beetle pollinated. The soft hairs on the seeds aid in their dispersal by the wind. Also known as Florida spiny pod. Purple milkweed vine (*M. biflora*) also has purplish-brown, star-shaped flowers; this species grows in Texas, Oklahoma, and New Mexico.

JOHN POLITES

ELEGANT BRODIAEA

Brodiaea elegans

Camas/Triteleia/Yucca family (Asparagaceae)

Description: Perennial, up to 25" tall. One to 3 narrow leaves arise from the base and wither prior to flowering. Blue to violet, funnel-shaped flower bears 6 recurved tepals. Along the flower's inner edge are 3 white antherless stamens that are pointed or tooth-topped and away from the 3 fertile stamens. Fruit is a capsule.

Bloom season: Spring to summer

Range/habitat: California and Oregon, mainly west of the Cascades, in meadows and open woodlands

Comments: *Brodiaea* honors James Brodie (1744–1824), a Scottish politician and botanist who studied cryptogamic plants such as algae and mosses. *Elegans* ("elegant") is in reference to the overall stature of the plant. The bulbs, or corms, were harvested and eaten by Native Americans.

COMMON CAMAS

Camassia quamash

Camas/Triteleia/Yucca family (Asparagaceae)

Description: Perennial. Grasslike basal leaves are about 1" wide and up to 2' long. The flowering stalk may bear numerous pale to deep blue (occasionally white) flowers, 1½" long. Flowers have 5 tepals that curve upward and 1 that curves downward. Fruit is an egg-shaped capsule.

Bloom season: Mid-spring to early summer

Range/habitat: Western North America in grassy plains or meadows

Comments: The common and scientific name is from a Native American name for the plant: *kamas* or *quamash*. Camas bulbs were an important food source for Indigenous peoples, and some camas fields were tended to by tribes. Bulbs were dug out of the ground and either ground into meal to store as cakes or roasted in earthen "ovens." Great camas (*C. leichtlinii*), a larger version of common camas, has a similar range.

JOHN POLITES

BLUE DICKS

Dipterostemon capitatus

Camas/Triteleia/Yucca family (Asparagaceae)

Description: Perennial, up to 30" tall. Leaves, 2–3, are 5"–20" long and keeled on the undersides. Cup-shaped flowers are borne in a cluster of 2–15 and have 6 blue to pink-purple (sometimes white) tepals. There are 6 deeply notched fertile stamens, 3 smaller ones on the outer tepals and 3 larger ones on the inner tepals; stamens are surrounded by 3 deeply notched appendages that lean in toward the center. Fruit is a capsule.

Bloom season: Winter to late spring

Range/habitat: Portions of western and southwestern United States into northwestern Mexico in grasslands, desert scrub, and open woodlands

Comments: *Dipterostemon* ("two-winged stamens") refers to the appendages on top of the anthers. *Capitatus* ("headlike") refers to the flower cluster arising at the end of the flowering stalk. The common name is a shortening of the genus name; also known as wild hyacinth. This plant is the sole species in this genus and was moved to this genus in 2021 after a taxonomic review.

ELEANOR DIETRICH

GEORGIA TICKSEED

Coreopsis nudata

Aster family (Asteraceae)

Description: Perennial, 3'–5' tall. Leaves are mostly basal, very reduced in size along the stems, and linear shaped up to 1" long. Flower heads have pink or purple ray flowers about 1" long that are notched at the tip surrounding a cluster of yellow disk flowers. Fruit is an oblong or linear seed.

Bloom season: Mid-spring to early summer

Range/habitat: Southeastern United States in moist locations such as bogs, wet prairies, savannas, moist woods, pine barrens, and ditches

Comments: *Coreopsis* is from the Greek *koris* ("bug") and *opsis* ("appearance"), which, along with the common name, refers to the shape of the seed, resembling a tick. *Nudata* ("without a leaf") refers to the leafless flowering stems. Bees are the dominant pollinators of these flowers, but other insects, such as butterflies and flies, are also attracted.

ELEANOR DIETRICH

SEASIDE DAISY

Erigeron glaucus
Aster family (Asteraceae)

Description: Perennial, mat-forming stems and a stout taproot, 2"–12" tall. Basal leaves are thick, inversely lance- to spatula-shaped, up to 4" long, and toothed along the upper margins. Leaves may also be covered with dense hairs or a white powdery coating. Flowering heads have numerous pinkish to white ray flowers (80–160) that surround a central cluster of yellow disk flowers. Heads are 1"–2" wide. Fruit is a seed with white hairs.

Bloom season: Spring and summer

Range/habitat: Western United States on coastal bluffs, beaches, and rocky headlands

Comments: *Erigeron* is from the Greek *eri* ("early") and *geron* ("old man") in reference to the white hairs on the seeds. *Glaucus* ("white coating") refers to white coloration on the leaves. The large flower surface attracts a variety of insect pollinators. These tough plants with somewhat succulent leaves often grow in areas exposed to surf spray.

STIFF-LEAVED ASTER

Ionactis linariifolia
Aster family (Asteraceae)

Description: Perennial, 6"–24" tall, often grows in clumps. Stalkless leaves are stiff, narrow, and ¾"–1½" long; arranged alternately along the stem or in "almost" whorls; undersides can be hairy. Flower heads are borne solitary or in clusters and are ¾"–1½" wide. Scalelike bracts in 4–5 series subtend these flower heads; linear to oblong bracts are green along the upper half and white along the lower half. There are 10–20 blue to violet ray flowers surrounding a center of yellow disk flowers that turn orange or reddish as they mature. Fruit is a bullet-shaped seed with hairs of varying lengths.

Bloom season: Late summer to fall

Range/habitat: Central and eastern North America in rocky outcrops, savannas, prairies, woodlands, and sandy locations

Comments: *Ionactis* is from the Greek *ion* ("violet") and *aktis* ("ray") in reference to the color of the ray flowers. *Linariifolia* ("leaves like *Linaria*") refers to the shape of these leaves, which resemble those in the Flax genus, *Linaria*. The flowers attract a variety of insect pollinators from flies to bees to butterflies. Another common name is flax-leaved aster, for the resemblance of its leaves to those of flax.

ERIC WATTS

RUSH SKELETON PLANT

Lygodesmia juncea
Aster family (Asteraceae)

Description: Perennial, 6"–18" tall. Stem and leaves contain milky latex. Lower leaves are ½"–2" long and grasslike; upper leaves are small and scalelike. A single flower is borne at the end of a flowering stalk. Flower heads are light pink to lavender (sometimes white), ½"–¾" wide, and composed of ray flowers. Linear-shaped rays have toothed tips. Bracts, which subtend the flower heads, are arranged in 2 rows of different sizes. Fruit is a seed with hairlike bristles.

Bloom season: Summer

Range/habitat: Western North America in dry, sandy, open locations, including plains, meadows, and deserts

Comments: *Lygodesmia* is from the Greek *lygos* ("pliant") and *desmia* ("bundle"), which refers to the thin stems. *Juncea* is from a Greek word meaning "rushlike," another reference to the stems. Rush-pink (*L. grandiflora*) and rose-rush (*L. aphylla*) are similar species that grow in the Southwest and Southeast, respectively. When the flower heads are closed, these plants may be easily overlooked.

TANSY-LEAF ASTER

Machaeranthera tanacetifolia
Aster family (Asteraceae)

Description: Annual or biennial, growing 6"–24" tall. Stem and leaves have sticky hairs. Pinnately divided leaves are ⅜"–2½" long, highly divided to the midrib; each segment ends with a soft, bristle tip. Flower heads are ⅜"–¾" wide and made of numerous disk flowers surrounded by 12–40 purplish ray flowers. The linear to lance-shaped bracts beneath the flower head have tips that curve either downward or outward. Fruit is a club-shaped seed with white hairs.

Bloom season: Late spring through fall

Range/habitat: Southwestern United States and northern Mexico in dry grasslands, meadows, and desert scrub

Comments: *Machaeranthera* ("sickle anther") refers to the shape of the anther. *Tanacetifolia* ("tansy-like leaves") refers to the highly dissected leaves. Hoary aster (*Dieteria canescens*), a variable and close relative, grows in the western United States. Also known as Tahoka daisy or tansy-aster.

ALPINE ASTER

Oreostemma alpigenum
Aster family (Asteraceae)

Description: Perennial, 2"–16" tall. Basal leaves are 1"–10" long, linear and untoothed, smooth to hairy. Solitary flower heads have 10–40 violet to lavender ray flowers surrounding a center of yellow disk flowers. Heads are subtended by a ring of green bracts, which are somewhat stiff. Fruit is a seed with hairy bristles.

Bloom season: Midsummer to early fall

Range/habitat: Western United States in mountain slopes, meadows, and rocky areas

Comments: *Oreostemma* is from *aros* ("mountain") and *stemma* ("a crown or garland"), referring to the profusion of flowers at higher elevations. *Alpigenum* ("alpine") refers to the plant's habitat. These plants attract butterflies as pollinators. Also known as mountain-crown or tundra aster.

LEAH BREITENSTINE

AMERICAN BASKET FLOWER

Plectocephalus americanus
Aster family (Asteraceae)

Description: Annual, 1'–5' tall; stem is ridged. Basal leaves are lance-shaped, alternate, stalkless, and generally absent when flowers bloom. Compound flower head has white to pinkish fertile inner flowers and purplish outer raylike flowers, which are sterile. Flower head may also have only disk flowers; the heads are 1½"–3¼" wide. Bracts below flower head in 8–10 overlapping rows, with tips bearing spinelike teeth. Barrel-shaped seeds have stiff bristles.

Bloom season: Spring to summer

Range/habitat: South-central United States and northern Mexico in sandy sites in open fields, pastures, and meadows

Comments: *Plectocephalus* is from the Greek *plectis* ("braided") and *kephale* ("head"), which refers to the weave of bracts that subtend the flower head; the common name also refers to this pattern. *Americanus* ("of America") refers to the plant's distribution. Native American peoples used different parts of the plant to treat various disorders such as venomous bites, liver disease, eye ailments, and burns.

ELEANOR DIETRICH

STOKES' ASTER
Stokesia laevis
Aster family (Asteraceae)

Description: Perennial, 1'–2' tall, with hairy stems. Stems arise from a basal rosette of lance- to egg-shaped evergreen leaves up to 6" long; stem leaves smaller and stalkless. Leaves may be dotted with tiny glands or smooth. Flower heads 2"–4" wide and composed of blue or bluish-purple disk flowers; outer flowers lobed and flare open, resembling ray flowers. Fruit is a seed with scaled bristles.

Bloom season: Late spring to midsummer

Range/habitat: Southeast United States in wet areas such as woodlands, bogs, coastal plains, and roadsides

Comments: *Stokesia* honors Jonathan Stokes (1755–1831), an English physician and botanist. *Laevis* ("smooth") refers to the leaf surface. This is the only species of this genus in the United States. The plants contain a monounsaturated vernolic acid, a key component of vernonia oil, which is extracted from the seeds and used in varnishes and paints. The flowers are pollinated by bees, butterflies, and wasps.

CHUCK TAGUE

NEW ENGLAND ASTER
Symphyotrichum novae-angliae
Aster family (Asteraceae)

Description: Perennial, 3'–6' tall. Stout stems branch and bear sticky hairs. Stem-clasping leaves with pointed tips are alternate, rough-textured, up to 4" long, and hairy. Flowering stems and bracts beneath the flowering head have sticky hairs; bracts are in 3–5 layers and green with a purplish tinge. Flowering heads, 1½" wide, bear over 30 purple to rose ray flowers surrounding a center of yellow disk flowers. Fruit is a seed with fine hairs.

Bloom season: Late summer to mid-fall

Range/habitat: Widely distributed from the Rocky Mountains to the East Coast in open fields, meadows, and seasonally moist to wet prairies

Comments: *Symphyotrichum* is from the Greek *symphysis* ("borne together or joined together") and *trichos* ("a hair"), referring to the ringlike arrangement of the pappus hairs. *Novae-angliae* ("of New England") refers to the plant's distribution. This and other leafy-stemmed asters are often called Michaelmas daisies. Native tribes in the Midwest made a tea from the roots to treat diarrhea or fevers. Introduced into European gardens in 1710, today this is a common garden flower.

BECKY

NEW YORK IRONWEED

Vernonia noveboracensis
Aster family (Asteraceae)

Description: Perennial, growing up to 6' in dense clusters. Alternate, lance-shaped leaves are 6"–8" long, saw-toothed along the margins, stiff, and pointed at the tip. Flower heads are 3"–4" wide, subtended by 4–7 rows of pointed bracts, and consist of 30–45 disk flowers (ray flowers are absent). Fruit is a rust-colored seed.

Bloom season: Late summer to early fall

Range/habitat: Eastern and southeastern United States in moist soils in thickets, floodplains, and stream banks

Comments: *Vernonia* honors William Vernon (c. 1666–1711), an English botanist and entomologist from Cambridge University who collected plants in Maryland in 1698. *Noveboracensis* ("of New York") refers to the type locality of the species. The common name refers to either the strong, ironlike strength of the stems; the rusty tinge on mature flowers; or the rust-colored seeds. These plants resemble Joe-Pye weed (*Eutrochium purpureum*) but have alternate leaves.

JOHN POLITES

PACIFIC HOUND'S TONGUE

Adelinia grande
Borage family (Boraginaceae)

Description: Perennial, 1'–3' tall. Basal leaves wide and elliptical with pointed tips. Leaves up to 7" long and hairy below. The flowering stalk bears cluster of ½"-wide, bluish trumpet-shaped flowers with 5 corolla lobes that spread outward from a white central tube. Fruit is a cluster of 4 round seeds covered with small hook-shaped bristles.

Bloom season: Spring

Range/habitat: Western North America in open woodlands and chaparral

Comments: *Adelinia* honors Adeline Etta Cohen (2014–) by her dad, Dr. James I. Cohen, a US botanist. *Grande* ("large") refers to the large leaves of this striking plant. Wild comfrey or blue hound's tongue (*Cynoglossum virginianum*) has light blue flowers and narrower leaves and grows in the southeastern and northeastern United States. The common name refers to the long, tongue-shaped leaves.

JACOB W. FRANK, NPS

ALPINE FORGET ME NOT

Eritrichium nanum
Borage family (Boraginaceae)

Description: Low mat-forming perennial, 2"–4" tall. The egg- to lance-shaped leaves are up to 2" long and hairy, forming a tuft of hairs at the tip. Funnel-shaped deep blue flowers are about ¼" wide with 5 petals and a yellow center ring. The flowers are borne in loose clusters on a short flowering stalk with small alternate leaves. Fruit is a nutlet.

Bloom season: Summer

Range/habitat: Circumpolar distribution in the Rocky Mountains and Alaska

Comments: *Eritrichium* is from the Greek *erion* ("wool") and *trichos* ("hair") and refers to the hairy leaves. *Nanum* ("little") refers to the stature of these plants, which, although low growing, are tough, long-lived perennials.

PATRICK ALEXANDER

STREAMBANK BLUEBELLS

Mertensia ciliata
Borage family (Boraginaceae)

Description: Perennial, 1'–4' tall, stems often arising in a cluster. Smooth, almost succulent blue-green leaves are oval to lance-shaped, pointed at the tip, and alternately arranged along the stem. The bluish flowers are arranged in loose pendant clusters. Individual tubular-shaped flowers are ⅓"–¾" long, flare open at the tip, and appear slightly crimped mid-flower; flower buds are pinkish, turning blue and then back to pinkish as the flowers mature. Inner portion of the corolla also has fine hairs. Fruit is a nutlet.

Bloom season: Late spring to late summer

Range/habitat: Western United States in moist hillsides, seeps, stream banks, and subalpine meadows

Comments: *Mertensia* is for Franz Karl Mertens (1764–1831), a German professor of botany. His son, Karl Heinrich Mertens (1796–1830) collected plants in the Northwest on Russian Count Fyodor Litke's globe-circling expedition of 1826–1829. *Ciliata* ("fringed") is for the small hairs that form a fringe along the leaf margins. Members of the *Mertensia* genus are often referred to as lungworts due to several species having been used to treat lung ailments.

STEVE R. TURNER

VIRGINIA BLUEBELLS

Mertensia virginica
Borage family (Boraginaceae)

Description: Perennial, 1½'–3' tall. Bluish-green leaves are oval shaped, about 4" long, pointed at the tip, and smooth. Trumpet- to bell-shaped flowers are 1" long and have narrow tubes that flare open at the tip; flower clusters arc downward. Bluish flowers may be pink in bud. Fruit is a nutlet with wrinkled surface.

Bloom season: Early to mid-spring

Range/habitat: Midwest to eastern United States in open woods, stream banks, and moist meadows

Comments: *Mertensia* is for Franz Karl Mertens (1764–1831), a German professor of botany. *Virginica* ("of Virginia") refers to the type locality of this species. Sea lungwort (*M. maritima*) is a low-growing relative that grows on beaches in northeastern North America. Bluebells have flowers with poricidal anthers that often require the strong vibration bumblebees make to force the anthers to open and release the pollen (called buzz pollination).

ALEX HEYMAN

DESERT CANDLE

Caulanthus inflatus
Mustard family (Brassicaceae)

Description: Annual, stems to 20"–30" tall. Thick stems are yellowish green, succulent, and inflated. Elliptical to lance-shaped leaves are 1"–3½" long, alternate, and clasp the stem with earlike lobes. The flowering stalk arises from a basal rosette of leaves and resembles a candle. Numerous ½"-long flowers have purple sepals that turn white as the flower matures; the 4 reddish-purple petals fade to white as well. Fruit is a 2"- to 4"-long narrow pod (silique).

Bloom season: Spring

Range/habitat: Western United States in open, sandy sites or rocky slopes

Comments: *Caulanthus* is from the Greek *kaulos* ("stem") and *anthos* ("flower"), which refers to cauliflower, since some species in the genus are used like that vegetable. *Inflatus* ("inflated") refers to the swollen stems. The tapered flowering stalk gives this plant its common name.

STEVE R. TURNER

TALL BELLFLOWER

Campanula americana

Harebell family (Campanulaceae)

Description: Annual or biennial, 1'–6' tall. Stems are mostly unbranched or with few side stems, slightly grooved, and rounded in cross section. Large, alternately arranged leaves are elliptical to egg shaped, up to 6" long, tapered at the tip, and with toothed margins. Flowers are borne along an elongated terminal stem or shorter side stems. The pale blue to violet star-shaped flowers are about 1" wide with 5 spreading lobes, a white center nectar ring, and a protruding style that curves downward. Fruit is a 5-angled, flat-topped capsule.

Bloom season: Early summer to fall

Range/habitat: Eastern half of United States in moist thickets and woodlands

Comments: *Campanula* ("small bell") refers to the modified bell shape of the flower. *Americana* ("of America") distinguishes this plant from European bellflower. Long-tongued bees and butterflies are the primary pollinators of tall bellflower. Seeds that germinate in the fall produce annual plants; seeds that germinate in spring produce biennials. Also known as American bellflower.

STEVE R. TURNER

HAREBELL

Campanula rotundifolia

Harebell family (Campanulaceae)

Description: Perennial, 6"–24" tall. Weak stems may be upright or arching. Alternate leaves are about 1" long, linear shaped, with entire margins; basal leaves have long stems and round to heart-shaped blades. Bluish-purple, bell-shaped flowers are 5-lobed, ½"–1¼" long, bearing 5 stamens and a 3-forked pistil. Petals are united about near their base. Fruit is a capsule.

Bloom season: Late spring to late summer

Range/habitat: Northern United States and Rocky Mountains in dry grasslands, meadows, rocky outcrops, and woodlands

Comments: *Campanula* ("small bell") refers to the shape of the flowers. *Rotundifolia* ("round leaf") refers to the shape of the leaves. Bees are common pollinators of these flowers. Olympic bellflower (*C. piperi*) is a low-growing relative found in the Olympic Mountains in Washington State. The common name may refer to the plants' growing in habitats with hares. Also known as bluebells-of-Scotland.

JANEL JOHNSON, NEVADA DIVISION OF NATURAL HERITAGE

BACH'S CALICOFLOWER

Downingia bacigalupii
Harebell family (Campanulaceae)

Description: Low-growing annual, 4"–20" tall, often in abundance. Diamond-shaped leaves are opposite/alternate and ¼"–¾" long. The 2-lipped bluish flower is ¼"–½" wide and has 2 long upper lobes that curl or project straight back. The lower lobes are fused into one 3-toothed lobe with 2 yellow spots rimmed with white. Both lobes bear dark blue veins. Fruit is a capsule.

Bloom season: Mid-spring to midsummer

Range/habitat: Western North America in meadows, floodplains, and seasonally wet areas

Comments: *Downingia* honors Andrew Jackson Downing (1815–1852), an American horticulturalist and editor of *The Horticulturist* (1846–1852). Downing is often considered the first superb landscape architect in America and had a great influence on architect Frederick Law Olmstead. *Bacigalupii* honors Rimo Bacigalupi (1901–1996), an American botanist and Jepson Herbarium curator at the University of California, Berkeley. The common name refers to the calico pattern of the spotted blue petals and Rimo, who was affectionately known as "Ba(t)ch."

KATIE BYERLY

GREAT BLUE LOBELIA

Lobelia siphilitica
Harebell family (Campanulaceae)

Description: Clump-forming perennial, 1'–4' tall, with square and sometimes slightly hairy stems. Lance-shaped leaves are up to 5" long, opposite, with finely toothed margins. Two-lipped pale to dark blue flowers are tubular shaped. The pale-striped flowers arise from leaf axils and have a 2-lobed upper lip and more-prominent 3-lobed lower lip, which has 2 prominent white bumps. Fruit is a capsule

Bloom season: Mid- to late summer

Range/habitat: Central and northeastern United States in wet meadows, along stream banks, swamps, and moist woodlands

Comments: *Lobelia* honors the French botanist and physician Matthias de l'Obel (1538–1616), who, with Pierre Pena, wrote *Stirpium Adversaria Nova* (1570), which detailed a new plant-classification system based on leaves not flower structure. *Siphilitica* refers to the plant's medicinal value for treating venereal disease.

STEVE R. TURNER

WHITEMOUTH DAYFLOWER

Commelina erecta
Spiderwort family (Commelinaceae)

Description: Perennial, to 4' tall. Linear to lance-shaped leaves have reddish edges, are 2"–6" long, taper to a tip, and have a rounded base. Leaf base clasps the stem to form a ½"- to 1"-long sheath that has a pair of rounded lobes projecting from the sheath's tip. A folded heart-shaped bract opens to reveal the emerging blue flower, which has 2 upper blue petals and a smaller, lower white one with 3 rounded lobes. Of the 6 stamens, 3 are sterile. Fruit is a capsule containing 2 brown seeds.

Bloom season: Early spring to fall

Range/habitat: Widespread across much of the Midwest and Southeast of the United States and scattered across the Southwest to California in dry woods, scrublands, and open forests

Comments: Linnaeus derived the generic name *Commelina* to honor Dutch botanists Jan Commelijn (1629–1692) and his nephew Caspar (1668–1731). *Erecta* ("erect") refers to the plant's stature. The common name refers to the flowers, which open and last for only 1 day. The elder Commelijn, along with Joan Huydecoper, founded Hortus Botanicus in Amsterdam, the Netherlands, one of the oldest botanical gardens in the world.

ELEANOR DIETRICH

OHIO SPIDERWORT

Tradescantia ohiensis
Spiderwort family (Commelinaceae)

Description: Perennial, 2'–4' tall. Alternate, gray or blue-green grasslike leaves are wider at the base, clasp the stem at an acute angle, and are up to 15" long and 1" wide. Light blue to violet-blue (sometimes white) flowers are borne in clusters at the top of a flowering stalk and emerge from 2 leaflike bracts. Each flower is 1"–1¾" wide, with 3 petals of equal length, 6 stamens with fine colored hairs at their base, and hairless sepals. Fruit is a woody capsule.

Bloom season: Late winter to fall, depending on location

Range/habitat: Across eastern and central United States and southern Ontario, Canada, in prairies, roadsides, meadows, thickets, woodlands, and disturbed areas

Comments: *Tradescantia* honors English gardener and botanist John Tradescant the Younger (1608–1662), who served as head gardener to Charles I after his father, John Tradescant the Elder, passed away. Tradescant the Younger collected spiderworts from Virginia, which he brought back to England. *Ohiensis* ("of Ohio") refers to where the plant was first collected for science. The leaves and stems are edible if cooked. Also known as bluejacket.

ELEANOR DIETRICH

HAIRY CLUSTER-VINE

Jacquemontia tamnifolia
Morning-glory family (Convolvulaceae)

Description: Annual twining vine, 3'–15' long. Stems may be greenish or reddish purple. Egg- to heart-shaped leaves are about 2" long and slightly hairy on both sides. Funnel-shaped blue flowers with a white center arise from leaf axils and in dense clusters; leaflike bracts subtend these clusters. Each flower has 5 hairy sepals. Flowers open in the evening and close during the day. Fruit is a rounded capsule; clusters of capsules resemble hair "puffballs."

Bloom season: Summer to winter

Range/habitat: Southeastern United States, West Indies, and into South America in disturbed ground or waste places, and sandy riverbanks

Comments: *Jacquemontia* is for French botanist and geologist Victor Jacquemont (1801–1832), who collected plants in India and the Himalayas. *Tamnifolia* ("leaves like *Tamnus*") refers to the resemblance of these leaves to those in the *Tamnus* genus, which is in the Yam family (Dioscoreaceae). Also known as tie-vine.

WOOLLY LOCOWEED

Astragalus mollissimus
Pea family (Fabaceae)

Description: Perennial, 2"–34" tall. Compound leaves are up to 11" long with 15–35 hairy, elliptical to egg-shaped leaflets. Flower stalks are purplish, densely covered with hairs, and bear 7–20 pea-shaped flowers. Individual bluish-purple to bicolored flowers are ¾" long and have a hairy calyx with 5 pointed lobes. The egg-shaped seedpods are ½"–1" long and hairy.

Bloom season: Early spring to late summer

Range/habitat: Southwestern and central United States in desert soils in grasslands and pinyon-juniper woodlands

Comments: *Astragalus* ("anklebone") refers to the shape of the seedpods. *Mollissimus* ("most soft") refers to the dense hairy covering of the leaves and stems, which gives them a soft texture. Members of this genus contain locoine, an alkaloid that may cause livestock to "go loco" or die if they graze too much on the plants.

STEVE R. TURNER

PURSH'S LOCOWEED

Astragalus purshii
Pea family (Fabaceae)

Description: Low-growing, mat-forming perennial. Leaves and stems are covered with dense hairs. Compound leaves are up to 6" long and bear 7–19 small oval to rounded leaflets with a pointed tip. Small white to purplish pea-shaped flowers are borne in clusters of 1–11. The ½"- to ¾"-long flowers are 2-lipped. Fruit is a hairy seedpod.

Bloom season: Spring to summer

Range/habitat: Western North America in desert shrublands, open plains, and mountain slopes

Comments: *Astragalus* ("anklebone") refers to the shape of the seedpods. *Purshii* honors Frederick T. Pursh (1774–1820), a German botanist who worked on the Lewis and Clark plant collections and wrote *Flora Americae Septentrionalis*, a flora of North America, in 1814. Also known as woolly-pod milkvetch.

PURPLE PRAIRIE-CLOVER

Dalea purpurea
Pea family (Fabaceae)

Description: Perennial, stems up to 3' tall. Compound leaves are odd-pinnate, with 3–7 opposite linear leaflets about 1" long. Stems and leaves may be covered with hairs or smooth and are dotted with small glands. Dense cluster of tiny rose to dark purple flowers borne in cylindrical heads or spikes, 2¾" long. Individual pealike flowers are tiny. Fruit is a pod with 1–2 seeds.

Bloom season: Midsummer to late summer

Range/habitat: Central United States and Canada in prairies, grasslands, roadsides, and woodlands

Comments: *Dalea* honors Samuel Dale (1659–1739), a British apothecary, physician, and amateur plant collector. *Purpurea* ("purple") describes the flower color. Meriwether Lewis first collected this plant for science in 1804 during the Lewis and Clark Expedition. Purple prairie-clover is a larval host plant for the southern dogface butterfly (*Zerene cesonia*).

MARGARET MARTIN

SILKY BEACH PEA

Lathyrus littoralis
Pea family (Fabaceae)

Description: Perennial, with gray-green stems that grow along the ground or slightly upright. Tendrils are present but aren't used for climbing. Compound leaves have 4–8 hairy, oval to oblong leaflets. Pea-shaped flowers are 2-toned, with the upper 2 petals pink-purple and the lower 3 petals white. Fruit is a hairy, oval-shaped pod.

Bloom season: Mid-spring to midsummer

Range/habitat: Coastline from British Columbia to southern California in dunes and sandy beaches

Comments: *Lathyrus* is from the Greek *laythros*, which is the name for peas. *Littoralis* ("shore") refers to the plant's common habitat. This beach pea is a host plant for various butterflies, including the painted lady, silver blue, western and eastern tailed blue, and orange sulphur.

BORREGOWILDFLOWERS.ORG

ARIZONA LUPINE

Lupinus arizonicus
Pea family (Fabaceae)

Description: Annual, up to 2' tall. Reddish-green stems have short and long hairs. Palmately compound leaves have 5–9 spatula- or inversely lance-shaped leaflets, which are ½"–1½" long and hairless on the upper side. Flowering stalks bear a spike of 20–50 magenta to deep pink pea-shaped flowers, ¼"–½" long. The upper or banner petal has a large white or yellow basal spot with red dots; this spot becomes magenta with age. Fruit is a 1"-long seedpod.

Bloom season: Late winter to late spring

Range/habitat: Southwestern United States from California to Arizona in open deserts and sandy washes

Comments: *Lupinus* is from the Latin *lupus* ("wolf"). Early peoples thought lupines "wolfed" nutrients from the soil; however, these and many other plants in the Pea family help fix nitrogen with their roots into a form that is useful for plants. *Arizonicus* ("of Arizona") refers to the plant's distributional range.

LARGE-LEAF LUPINE

Lupinus polyphyllus
Pea family (Fabaceae)

Description: Tall, up to 5' tall with stout stems. Compound leaves are palmately divided and have 9–17 elliptical to lance-shaped leaflets that are smooth above and may be slightly hairy below. Individual leaflets are 2"–5" long. Pea-shaped flowers are borne on elongated stalks in dense clusters and are bluish purple but sometimes pink, white, or 2-toned. The ½"-long flowers have a smaller upper petal that curls backward and is shorter than the 2 lateral petals. Fruit is a seedpod.

Bloom season: Mid-spring to midsummer

Range/habitat: Western North America in moist habitats such as stream banks, river edges, and wetlands

Comments: *Lupinus* is from the Latin *lupus* ("wolf"). Early peoples thought lupines "wolfed" nutrients from the soil; however, these and many other plants in the Pea family help fix nitrogen with their roots into a form that is useful for plants. *Polyphyllus* ("many leaves") refers to the numerous leaflets. This lupine has been horticulturally crossed with other lupines to form several common hybrids. Also called bigleaf lupine.

KATHY RIGALL

TEXAS BLUEBONNET

Lupinus subcarnosus
Pea family (Fabaceae)

Description: Annual, 6"–16" tall. Palmately compound leaf with 5 oblanceolate leaflets, each with a small pointed tip. The bluish, pea-shaped flowers are borne along an elongated stalk, and the upper petal has a white center that fades to a purplish color with age. The flowers are arranged in a looser arrangement than many lupines. Fruit is a seedpod.

Bloom season: Early to late spring

Range/habitat: Southern United States from Texas to Florida in sandy fields and roadsides.

Comments: *Lupinus* means "wolf" and refers to the belief that lupines "wolf" nutrients from the soil; however, these and many other plants in the Pea family help fix nitrogen with their roots into a form that is useful for plants. *Subcarnosus* ("having the beginnings of a keel") refers to the flowers. In 1901 this species was named the state flower of Texas, but that changed in 1971 to include all *Lupinus* species found in the state as the state flower. Also called sandyland bluebonnet for the plant's habitat preference and because the shape of the petals resembles the bonnet worn by pioneer women.

STEVE R. TURNER

PURPLE LOCOWEED

Oxytropis lambertii
Pea family (Fabaceae)

Description: Perennial, 4"–16" tall. Compound leaves arise from the base and are up to 6" long. Linear to oblong leaflets are ⅓"–1" long, between 7–19 per leaf, and covered in long silver hairs. Pea-shaped flowers are pink-purple to violet, ¾" long; upper petal is notched at the tip. The 2 lateral petals fold over a tapered beak-like keel. Flowers are arranged in an elongated cluster of up to 25 flowers per stalk. Fruit is a hairy, cylindrical seedpod.

Bloom season: Spring and summer

Range/habitat: Central Canada south through much of central United States in prairies and grasslands

Comments: *Oxytropis* ("sharp keel") refers to the flower. *Lambertii* honors British botanist Aylmer Lambert (1761–1842), who helped Frederick Pursh with the Lewis and Clark collection and wrote *A Description of the Genus Pinus*, which described many of the conifers David Douglas collected. A fungus in the *Embellisia* genus lives within this plant's cells during a portion of its life cycle. This fungus may produce a toxic alkaloid called swainsonine, which causes poisoning in livestock that forage on this locoweed.

WHITE-STEM FRASERA

Frasera albicaulis
Gentian family (Gentianaceae)

Description: Perennial, 10"–30" tall. Basal leaves linear, up to 6" long with white margins. Upper leaves smaller and arranged oppositely. Dense flower cluster borne at the end of an elongated stalk. Blue (sometimes greenish-white) flowers, ½" wide, with 4 lance-shaped, mottled petals. Petals have a hairlike gland at the base. Fruit is a small capsule.

Bloom season: Late spring to midsummer

Range/habitat: Western North America in meadows, shrub steppe, rocky areas, and open woods

Comments: *Frasera* honors John Fraser (1756–1811), a Scottish botanist and explorer who traveled extensively and collected plants in America, Russia, Canada, and Cuba in the eighteenth century. He introduced over 200 plants into European gardens. *Albicaulis* ("white-stemmed") refers to the edges of the leaves. Also known as elkweed.

STEVE R. TURNER

PLEATED GENTIAN

Gentiana affinis

Gentian family (Gentianaceae)

Description: Perennial, 6"–30" tall. Narrow, lance-shaped leaves are oppositely arranged, up to 1½" wide, with short fine hairs along the edge. Each pair of leaves arises from the stem at right angles from the pair below. Bluish-purple flowers bearing white or green spots are borne in leaf axils or at the end of a flowering stalk. Funnel-shaped flowers are 1"–1½" long and 5-lobed, with the lobes widely spreading and connected by pleat-like tissue. Fruit is a narrow capsule that bears numerous tiny seeds with narrow wings.

Bloom season: Mid-spring to late summer

Range/habitat: Western United States in moist meadows, prairies, and wet montane areas

Comments: *Gentiana* is from the Greek name for King Gentius, who is often cited as the discoverer of the medicinal properties of gentians native to his country. *Affinis* ("closely related" or "akin to") refers to this plant resembling another species of gentian. The white and green dots act as pollinator guides, leading insects into the flower.

STEVE R. TURNER

BOTTLE GENTIAN

Gentiana andrewsii

Gentian family (Gentianaceae)

Description: Perennial, often 1'–3' tall, from a stout taproot. Strap-like opposite leaves are ovate or broadly lance-shaped, up to 4½" long, and stemless; upper leaves arise in a whorl. Bottle-shaped flowers, 1"–1½" long, arise in the upper leaf axils or at the end of the stem. Flowers remain closed even at maturity and appear "wrinkled" due to longitudinal ridges. Fruit is a small seed.

Bloom season: Late summer to early fall

Range/habitat: North-central United States and Canada to northeastern United States in moist locations such as fens, bogs, wetlands, and thickets

Comments: *Gentiana* is from the Greek name for King Gentius (reign c. 180–168 BCE), who was reputed to have discovered the medicinal virtues of the root of the yellow gentian, or bitterwort (*G. lutea*), from which a bitters tonic is still made. *Andrewsii* honors English botanical artist Henry C. Andrews (1794–1830). According to ancient Roman naturalist Pliny, King Gentius of Illyria found that the roots were useful as an emetic, cathartic, and tonic. The common name also comes from this king. The small seeds may be transported by the wind or water.

KATIE BYERLY

GREATER-FRINGED GENTIAN

Gentianopsis crinita
Gentian family (Gentianaceae)

Description: Biennial, 1'–3' tall. Slightly succulent leaves are 1"–2½" long, egg- to narrowly lance-shaped, and oppositely arranged. Solitary flowers are borne on leafless stalks up to 4" tall. The 4-lobed, tubular blue flowers have fringed petals subtended by 2 pairs of green keeled sepals. Fruit is an elongated capsule.

Bloom season: Late summer and fall

Range/habitat: North-central and northeastern United States and Canada in bogs, fens, wooded swamps, ditches, and open, moist sandy areas

Comments: Members of the *Gentianopsis* genus resemble *Gentiana* but lack the connective tissue between the flower lobes. According to ancient Roman naturalist Pliny, King Gentius of Illyria found that the roots of gentians were useful as an emetic, cathartic, and tonic. *Crinita* ("long, weak hairs") refers to the shape of the hairs.

STEVE R. TURNER

WILD GERANIUM

Geranium maculatum
Geranium family (Geraniaceae)

Description: Perennial, clump forming; stems to 2' tall. Palmately divided leaves are up to 6" wide and have 3–7 deep lobes, further divided into 3–5 sections that are cleft and toothed along the upper half. Saucer-shaped flowers are purplish pink to lavender, 1¼" wide, 5-lobed, with nectar guide lines. Fruit is a capsule.

Bloom season: Mid-spring to late spring

Range/habitat: Eastern North America in dry woodlands

Comments: *Geranium* is from the Greek *geranos* ("crane"), referring to when the mature capsule opens and each seed is attached to a long, beak-like column, formed by the fused styles, that resembles a crane's head and beak. *Maculatum* ("spotted") refers to the spotting on the petals. The plant was used medicinally as a poultice to stop bleeding. Also known as spotted crane's-bill.

JUDY PERKINS

BALLHEAD WATERLEAF

Hydrophyllum capitatum
Waterleaf family (Hydrophyllaceae)

Description: Perennial, 4"–20" tall. Compound leaves are mostly basal and borne on long stems. Deeply divided 7–11 leaflets are again lobed or divided. The small, purplish-blue to white flowers arise in rounded, 1"- to 2"-wide clusters sometimes hidden by the leaves. Individual flowers are small, 5-lobed, with stamens that protrude above the corolla. Fruit is a capsule.

Bloom season: Early spring to midsummer

Range/habitat: Much of western North America from British Columbia to Utah in woodlands, thickets, and meadows

Comments: *Hydrophyllum* ("water leaf") may refer to the spots on the leaves that resemble water stains. *Capitatum* ("growing in a dense head") refers to the flower clusters. Native Americans and early pioneers boiled and ate the leaves as greens.

BABY BLUE-EYES

Nemophila menziesii
Waterleaf family (Hydrophyllaceae)

Description: Annual, low growing, 3"–6" tall. Stems may trail along the ground. Lower leaves opposite, pinnately lobed with 5–13 lobes, up to 2" long; each lobe may be entire or toothed. Flowers borne solitary, cup-shaped, up to 1½" wide; 5 blue petals with a white center, although some flowers are white overall with blue veins and black dots around the center. Fruit is a small, hairy capsule.

Bloom season: Spring to early summer

Range/habitat: Western from Oregon to Baja California in dry chaparral, grasslands, desert washes, and montane meadows

Comments: *Nemophila* is from the Greek *nem*os ("wooded pasture or glade") and *phileo* ("to love") in reference to the habitat some species in the genus prefer. *Menziesii* honors Archibald Menzies (1754–1842), a Scottish botanist and surgeon who explored the Pacific Northwest with George Vancouver during an expedition in 1790. Widely used in ornamental plantings. Five spot (*N. maculata*) is closely related but has white cup-shaped flowers with purple spots on the petal tips.

STEVE R. TURNER

FERNLEAF PHACELIA

Phacelia bipinnatifida
Waterleaf family (Hydrophyllaceae)

Description: Biennial, 1'–3' tall, loosely branched. Stems have glandular hairs. First-season leaves are up to 5" long and divided into 3–5 (rarely 7) deep, coarsely lobed fernlike leaves. The flowering stalk arises in the second season and bears 1"-wide, shallow disk-shaped flowers with 5 sepals and 5 lavender to blue-violet petals. Fruit is an oval capsule.

Bloom season: Mid-spring to early summer

Range/habitat: Southeastern United States in open woods, ravines, and rocky slopes

Comments: *Phacelia* ("fascicle") describes the tightly clustered arrangement of the flowers. *Bipinnatifida* ("twice pinnately cut") refers to the highly dissected leaves. Miami mist (*P. purshii*) also grows in this region and has fringed petals. Bees are the more-common pollinators of these flowers, which also attract butterflies and wasps. Also called forest phacelia for its habitat preference.

SCORPIONWEED

Phacelia crenulata
Waterleaf family (Hydrophyllaceae)

Description: Annual, 2"–32" tall; may grow in profusion. Basal leaves are strap-shaped to elliptical, with margins deeply lobed or wavy. Leaves are ¾"–4¾" long. Sticky, glandular hairs mix with nonglandular hairs along the leaves and stems. Flowers grow along an elongated axis that curls like a scorpion's tail. Blue-violet to purple flowers are ¼" long and bell-shaped. Stamens and style protrude above the flowers. Fruit is a seed.

Bloom season: Mid-spring to midsummer

Range/habitat: Southwestern United States and northern Mexico in grasslands, desert shrub, mountain brush, open woodlands, and rocky areas

Comments: *Phacelia* is from the Greek *phakelos* ("fasicile"), which refers to the clustered flowers that are arranged in a "scorpioid cyme." *Crenulata* ("shallow, rounded teeth") refers to the leaf margins. The plants emit a strong odor, which may be to deter herbivores.

THREAD-LEAF PHACELIA

Phacelia linearis
Waterleaf family (Hydrophyllaceae)

Description: Annual, grows 5"–20" tall. Basal leaves are narrow; larger upper leaves may be split into 1–4 narrow segments. Showy, light blue flowers are ½"–1" wide and have 5 petals. Flower buds are arranged in a coiled pattern. Fruit is a rough-textured seed.

Bloom season: Late spring and early summer

Range/habitat: Western North America in plains, meadows, sagebrush steppe, and open woodlands

Comments: *Phacelia* ("fascicle") describes the tightly clustered arrangement of the flowers. *Linearis* ("linear") refers to the shape of the leaves. The dish-like flowers attract bees, flies, and beetles as the primary pollinators. Some years, these plants bloom in profusion.

SILKY PHACELIA

Phacelia sericea
Waterleaf family (Hydrophyllaceae)

Description: Perennial, stems 4"–20" tall and hairy. Lance-shaped basal leaves are pinnately lobed and covered with silky hairs. Leaves may be divided several times; individual segments have rounded lobes. Flowering stalks bear a thick cluster of bluish-purple, 5-lobed flowers that are bell-shaped with exserted stamens that are nearly twice as long as the corolla. Fruit is a 2-chambered capsule.

Bloom season: Late spring through summer

Range/habitat: Western North America in subalpine or alpine open woodlands and rocky slopes

Comments: *Phacelia* ("fascicle") describes the tightly clustered arrangement of the flowers. *Sericea* ("silky") refers to the abundant fine hairs on the stem and leaves. Many *Phacelia* species are annuals or biennials, but silky phacelia is a long-lived perennial. Studies conducted in British Columbia concluded that these plants can accumulate trace amounts of minerals, such as gold, in their tissues. The flowers attract mostly bees as pollinators.

TOM LEBSACK

PROPELLER-FLOWER

Alophia drummondii
Iris family (Iridaceae)

Description: Perennial, 6"–24" tall, from a bulb. Leaves are linear-lanceolate, 1'–1½' long, and folded down the middle (pleated). Cup-shaped flowers have 6 tepals; outer ones are broader and larger than the inner 3 boat-shaped or crimped ones. Flower is about 1½" wide. Tepals are purplish with yellow centers spotted with dark purple or reddish brown. Fruit is a capsule.

Bloom season: Mid-spring to midsummer

Range/habitat: Southern United States from Texas to Oklahoma and south into Mexico in sandy prairies, forest edges, and woodlands

Comments: *Alophia* is from the Greek *a* ("without") and *lophus* ("a crest"), referring to the lack of a style crest. *Drummondii* honors Thomas Drummond (c. 1790–1835), a Scottish naturalist who traveled to the United States in 1830 to collect plants. Drummond ended up in the Texas region in 1833 and collected extensively in that region for more than 20 months. He died in Havana, Cuba, in 1835 while collecting plants on that island.

WESTERN BLUE FLAG

Iris missouriensis
Iris family (Iridaceae)

Description: Perennial, 1'–3' tall. Strap-like leaves are less than ½" wide and 7¾"–15" long. The leaves are generally longer than the flowering stalks. The pale blue flowers are roughly 3" wide. Flowers have 3 wide sepals lined with purple streaks and a yellow base, 3 upright smaller petals, and 3 petallike styles that are aligned with the sepals. Fruit is a large capsule.

Bloom season: Late spring to summer

Range/habitat: Western North America in wet meadows, riverbanks, and marshy areas

Comments: Iris was the Greek goddess of the rainbow, often personified by the colors of the rainbow. *Iris* refers to the multihued colors of the petals. *Missouriensis* ("of the Missouri River") indicates where this species was first collected for science, in 1806 during the Lewis and Clark Expedition. Native Americans pounded the roots of western blue flag to treat toothaches; the strong fibers in the stems and leaves were used in making cordage.

JENNIFER AITKENS

NORTHERN BLUE FLAG

Iris versicolor
Iris family (Iridaceae)

Description: Perennial, 2'–3' tall and clump forming. Swordlike leaves are folded along the midvein, bluish green, and up to 1" wide and 2'–4' long. Flowering stalks rise up to about 30" tall and bear 3–5 violet-blue flowers that are about 4" wide. The sepals have a central yellow spot surrounded by white and show purplish veins and have greenish-yellow hairs (the signal) at the throat. Fruit is a capsule.

Bloom season: Late spring to midsummer

Range/habitat: Eastern North America south to Virginia in meadows, marshes, and stream banks

Comments: Iris was the Greek goddess of the rainbow, and *Iris* refers to the multihued colors of the petals. *Versicolor* ("variously colored") refers to the flowers. "Flag" is from middle English *flagge* ("rush or reed") and refers to the flagged irises growing in the same habitat frequented by rushes or reeds. The roots and leaves contain a toxic glycoside called iridin. Blue flag is the provincial flower of Quebec, and the iris (any blue to purplish one) is considered the state flower of Tennessee.

TOM LEBSACK

PRAIRIE-CELESTIAL

Nemastylis geminiflora
Iris family (Iridaceae)

Description: Perennial from a ½"- to 1½"-long bulb; plants are 5"–26" tall. Grasslike leaves, 1–4, are 5"–18" long and pleated along the midrib, showing ribs and furrows. Saucer-shaped blue flowers are 1½"–3" wide and made up of 6 tepals—outer ones are larger—and 3 stamens. Flowers are borne in small clusters of 1–3 flowers on the ends of slender branches. Fruit is an egg-shaped capsule that is flat on the top and tapers to a point at the base.

Bloom season: Early to late spring

Range/habitat: South-central United States in dry prairies and oak woodlands

Comments: *Nemastylis* is from the Greek *nema* ("thread") and *stylo* ("pillar"), in reference to the upright and threadlike styles. *Geminiflora* ("twin flowered") refers to the arrangement of two flowers arising on separate stalks from a common point. The flowers last only part of a day. Also known as prairie pleatleaf.

STEVE R. TURNER

NARROW-LEAF BLUE-EYED-GRASS

Sisyrinchium angustifolium
Iris family (Iridaceae)

Description: Perennial, 4"–18" tall and clump forming. Grasslike leaves 12"–20" long. Branched flowering stems are flattened, but with 2 winged edges. Blue-violet flowers that darken toward the yellow centers are ½" wide and arise from the stem in leaf axils that are subtended by a leafy bract. Flowers have 6 tepals; each tepal ends in a sharply pointed tip. Fruit is a capsule.

Bloom season: Spring to midsummer

Range/habitat: Eastern North America in moist meadows and open woodlands

Comments: *Sisyrinchium* is for a genus of plants that Carl Linnaeus named in 1753 and is derived from a Greek word for another plant, the Barbary nut iris (*Iris sisyrinchium*). *Angustifolium* ("narrow leaved") refers to the grasslike leaves. Idaho blue-eyed-grass (*S. idahoensis*) grows in western North America and has similar blue flowers that arise on unbranched stems.

STEVE R. TURNER

OHIO HORSEMINT

Blephilia ciliata
Mint family (Lamiaceae)

Description: Perennial, 1'–3' tall and clump forming. Square stems are mostly unbranched, have short appressed hairs, and bear opposite lance-shaped leaves up to 3½" long and toothed along the margins. Leaves are covered with fine hairs. Bluish-purple to white 2-lipped flowers arise in tiers along the stem. The 2"- to 3"-wide globular flower clusters are subtended by a whorl of leafless bracts fringed with hairs. Each flower has 2 projecting stamens, small purple spots on the lower lip, and fine hairs on the back of the flower. Fruit is a seed.

Bloom season: Late spring to summer

Range/habitat: Eastern North America in open areas, thickets, dry meadows, and occasionally woods

Comments: *Blephilia* ("eyelash") refers to the hairs that fringe the leaf bracts resembling eyelashes. *Ciliata* ("hairlike appendages") refers to the fine hairs on this plant. The foliage of this mint has minimal aroma but attracts numerous types of pollinators. Hairy wood-mint (*B. hirsuta*) also grows in this region but has longer spreading hairs and grows more in woodland habitats. Ohio horsemint is also called downy wood-mint.

KATIE BYERLY

BLUE SAGE

Salvia azurea

Mint family (Lamiaceae)

Description: Clump-forming perennial, 3'–5' tall. Stiff, square stems bear linear to lance-shaped 3"- to 5"-long leaves that are opposite, pointed at the tip, and smooth or toothed along the upper half of the leaf margin. Whorls of 2-lipped, ¼"- to 1"-long, bluish-purple flowers have a prominent lower lip bearing a large white spot. Tubular calyx is hairy. Fruit is a seed.

Bloom season: Summer and fall

Range/habitat: Central and eastern North America in open areas, prairies, and woodlands

Comments: *Salvia* ("to save or heal") is a reference to the medicinal properties of certain species in this genus. *Azurea* ("sky blue") refers to the flower color. The larger lip provides a landing platform for bees, which crawl beneath the upper lip toward the nectar glands before their back gets dusted with pollen from a fertile anther. Also known as pitcher sage, for Dr. Zina Pitcher (1797–1872), a physician, educator, president of the American Medical Association, and botanist who collected plants in the Great Lakes region.

BORREGOWILDFLOWERS.ORG

CHIA

Salvia columbariae

Mint family (Lamiaceae)

Description: Annual, 4"–28" tall. Deeply divided leaves are mostly basal, up to 4" long, with short stiff hairs and irregularly rounded lobes. Uneven surface of the leaves makes them appear wrinkled. Square and slightly hairy flowering stalk bears several clusters of bluish 2-lipped flowers, which are subtended by purplish spiny bracts that make up the calyx. Flower's lower lip is larger than the upper one, and the 2 stamens barely protrude beyond the flower. Fruit is a nutlet.

Bloom season: Spring

Range/habitat: Western United States from California to New Mexico and south into Mexico in dry, open areas, desert washes, and chaparral

Comments: *Salvia* ("to save or heal") is a reference to the medicinal properties of certain species in this genus. The derivation of *columbariae* is unclear. The Aztec *chian* ("oily") is the source of the common name: chia. The nutty-flavored seed was very important to Indigenous peoples and was eaten raw or cooked. The seeds are also a key food for mourning doves. Also known as California sage.

STEVE R. TURNER

DOWNY SKULLCAP

Scutellaria incana
Mint family (Lamiaceae)

Description: Perennial, 2'–3' tall, with fine white hairs on the square stems. Leaves are opposite, lance- to egg-shaped, up to 3" long, toothed along the margins, and with rounded bases. Small purplish-blue, 2-lipped flowers arise in dense clusters along an elongated flowering stalk. Tubular corolla is purplish blue, about ¾" long, and covered with fine hairs on the top. Upper lip is hoodlike; the lower lip is broader and bears a large white patch on the corolla's throat. The ¼"-long calyx has a unique projection on its topside. Fruit is an odd-shaped capsule.

Bloom season: Summer

Range/habitat: Southeast and central United States in thickets, stream banks, and woods

Comments: *Scutellaria* ("a small dish or saucer") refers to the calyx's shape when the flowers fade. *Incana* ("hoary" or "very gray") refers to the coloration of the stem and leaves, which are covered in fine hairs. The common name skullcap refers to the cap- or hoodlike shape of the upper petal and the shape of a bulge on the calyx resembling a helmet worn by warriors during the Middle Ages. Also called hoary skullcap due to the fine white hairs.

STEVE R. TURNER

MAD-DOG SKULLCAP

Scutellaria lateriflora
Mint family (Lamiaceae)

Description: Perennial, 30"–40" tall. Stems are square and leaves are opposite, egg- to lance-shaped, up to 3" long, and toothed along the margins. Flowers are small, ⅓" long, and arranged along 1-sided flowering stalks that mostly arise from leaf axils. The 2-lipped flowers are blue to pink or white; often the upper lip is bluer and the lower lip is white. Calyx has 2 lobes; upper lobe has a saucer-shaped projection. Fruit is a capsule.

Bloom season: Summer and early fall

Range/habitat: Mostly central and eastern North America, but with scattered populations in the west from British Columbia to California in moist areas such as wetlands, marshes, and meadows

Comments: *Scutellaria* ("a small dish or saucer") refers to the calyx's shape when the flowers fade. *Lateriflora* ("side flowering") is in reference to the one-sided flowering stalks. This and other skullcaps contain flavonoids in the leaves, which are used medicinally as a sleep aid and antispasmodic. The common name skullcap refers to the cap- or hoodlike shape of the upper petal and the capsules resembling helmets worn by warriors during the Middle Ages. The common name mad-dog comes from the false belief that this plant could cure rabies in dogs.

STEVE R. TURNER

FORKED BLUECURLS

Trichostema dichotomum
Mint family (Lamiaceae)

Description: Annual, 6"–24" tall. Upper stems and side branches bear sticky hairs. Leaves are opposite, entire, narrowly elliptical, and up to 2½" wide. Clusters of 2–7 flowers arise on opposite sides of the main stem and are subtended by 2 opposite leaves. Flowers are 2-lipped bearing 5 petals that are violet-blue in color, with the lower lobe whitish (sometimes blue) and bearing blue to purple spots. Lower lip is the largest petal, tongue-like, and bends downward. The 4 bluish-purple stamens extend beyond the flower's throat in an arching or curled pattern along with the style. Fruit is a cluster of 4 small nutlets.

Bloom season: Summer and fall

Range/habitat: Eastern North America, central United States, and the Bahamas in sandhills, pine woodlands, dry oak woodlands, and disturbed areas

Comments: *Trichostema* ("hairlike stamens") in reference to the very narrow stamens. *Dichotomum* ("forked stigma") refers to the shape of the stigma, which forks at the top. The common name refers to the arching stamens and the forked style. The leaves give off a lemony scent when crushed.

BORREGOWILDFLOWERS.ORG

DESERT CHRISTMAS-TREE

Pholisma arenarium
Lennoa family (Lennoaceae)

Description: Mat-forming, fleshy perennial, 3"–8" tall, with the stems buried underground. The aboveground portion may be capsule- or cone-shaped. Plants lack chlorophyll, so the plant is brownish gray or whitish overall. Pointed leaves are scalelike, alternate, with glandular hairs appressed to the body. Funnel-shaped flowers are ½" wide with 4–10 lavender to purplish lobes edged with white arising in clusters between the leaves. Fruit is a fleshy capsule.

Bloom season: Spring to midsummer

Range/habitat: Southern California to southern Arizona and northwestern Mexico in sandy washes, desert chaparral, and coastal dunes

Comments: *Pholisma* is from the Greek *pholis* ("scales") in reference to the scalelike leaves. *Arenarium* ("sand loving") refers to the type of habitat this plant grows in. This plant is parasitic on burrobush, rabbitbrush, yerba santa, and other shrubs. The plant sends out lateral roots beneath the soil to contact the host's roots. The common name refers to the plant's conical, Christmas tree shape. Sand food (*P. sonorae*) is similar, but either mushroom shaped to rounded where the edible stems are buried by the sand.

ELEANOR DIETRICH

BLUE BUTTERWORT

Pinguicula caerulea

Bladderwort family (Lentibulariaceae)

Description: Perennial, carnivorous plant, 4"–12" tall. Yellowish-green basal leaves, up to 3½" long, have tiny hairs that emit a sticky substance to trap insects. Lavender-blue flowers with 5 notched petals streaked with purple are borne at the end of a leafless, hairy stalk. Flowers are about 1" long and have a spur; sepals also have glandular hairs. Fruit is a capsule.

Bloom season: Midwinter to early summer

Range/habitat: Southeastern United States in moist sandy pinewoods or bogs

Comments: *Pinguicula* is from the Latin *pinguis* ("fat") and refers to the greasy texture of the leaf surface. *Caerulea* ("sky" or "sea") is in reference to the sky-blue color of the flowers. When an insect becomes stuck to the hairs, the leaves fold over and the digestive process begins. The common name refers to the flower color and the historical use of the leaves to curdle milk.

ELEANOR DIETRICH

PURPLE BLADDERWORT

Utricularia purpurea

Bladderwort family (Lentibulariaceae)

Description: Perennial, carnivorous plant, 1"–6" tall. Submerged stems support whorls of 5–7 leaves, each with a bladder on the tip. A leafless flowering stalk rises 2"–5" above the water surface and bears 1–4 pinkish-purple flowers with a yellow spot and short spur. The ⅓"- to ½"-long flowers have a larger lower lip that becomes 3-lobed and is white at the base; the upper lip is shorter. Fruit is a round capsule.

Bloom season: Late spring to fall

Range/habitat: Eastern North America in ponds, lakes, and shallow waters

Comments: *Utricularia* has various meanings but may be derived from a word meaning "leather flask" or "bagpipe," referring to the shape of the bladders. *Purpurea* ("purple-like") refers to the color of the flowers. Bladderworts do not have roots but are floating plants that trap insects. The leaf bladders are mainly for capturing and digesting zooplankton. New research indicates that the bladders contain microorganisms such as bacteria, algae, and diatoms that may have a symbiotic relationship with their hosts.

STEVE R. TURNER

SAGEBRUSH MARIPOSA LILY

Calochortus macrocarpus
Lily family (Liliaceae)

Description: Perennial, 8"–21" tall, from a bulb. The 1–3 grasslike leaves are thin, grow up to about 4" long, and often wither prior to flowering. One to 5 flowers are borne on each plant, and the flowers are about 2" wide. Flowers are lavender, purple, or white, with 3 narrow, pointed sepals that are longer than the 3 broad petals. A green stripe extends down the petals to the sepals, and reddish-purple nectar lines extend on the lower, inner portion of the petals. The light-colored center bears numerous yellow hairs. Fruit is a capsule.

Bloom season: Late spring to midsummer

Range/habitat: Western North America in dry meadows, sagebrush steppe, or ponderosa pine woodlands

Comments: *Calochortus* ("beautiful grass") refers to the overall beauty of this plant with the grass-like leaves. *Macrocarpus* ("large fruited") refers to the large seed capsules, which split open and release seeds when mature. Native Americans harvested the onion-like bulbs for food.

WESTERN BLUE FLAX

Linum lewisii
Flax family (Linaceae)

Description: Perennial, 1'–2' tall. Thin stems bear linear to lance-shaped leaves that are 1"–2" long. The 1"-wide blue flowers have 5 petals streaked with purple nectar guides and 5 styles that protrude beyond the stamens. Flower center is yellowish. Fruit is a small, round capsule.

Bloom season: Mid-spring to midsummer

Range/habitat: Western North America in damp or dry meadows, and woodlands

Comments: *Linum* is the Latin name for flax. *Lewisii* honors Meriwether Lewis (1774–1809), co-leader of the Corps of Discovery Expedition, better known as the Lewis and Clark Expedition. Lewis collected and described blue flax on his western expedition, noting its perennial form and potential commercial properties. Some western Native tribes used flax fibers to make cordage for nets and snares. Fiber from the cultivated common flax (*L. usitatissimum*) has been woven into linen for more than 300 years.

STEVE R. TURNER

WINGED LOOSESTRIFE

Lythrum alatum
Loosestrife family (Lythraceae)

Description: Perennial, 1'–4' tall, with square stems that have slightly raised ridges or wings along the edges. Sword-shaped leaves are opposite near the base and alternate higher up on the stem. Leaves are narrow at the tip, have smooth edges, and are up to 3½" long. Lavender flowers arise in pairs or singly from upper leaf axils. Calyx is 5-lobed, each lobe ending with a pointed tip. Fruit is an elongated capsule.

Bloom season: Early summer to early fall

Range/habitat: Central and eastern United States in wet meadows, ponds, bogs, fens, stream banks, river edges, and ditches

Comments: *Lythrum* is from the Greek *lythron* ("blood"), which refers to either the plant's styptic properties or the color of the flowers. *Alatum* ("winged") is in reference to the raised ridges along the stems. The common name loosestrife is derived from the ancient use of garlands of these flowers being draped over the yoke of oxen to reduce the distraction of flying insects so the oxen would calmly plow a field.

CHUCK TAGUE

ALLIGATOR FLAG

Thalia geniculata
Arrowroot family (Marantaceae)

Description: Clump-forming aquatic perennial that can grow 6'–9' tall or greater. Leaves are borne on long, thick stalks, broadly lance-shaped, and up to 30" long. The underside midrib of the leaf is prominent. Flowering stalks are long and branched, with the flowers clustered near the tips of the zigzag-shaped branches. The unique flowers are enclosed within green to purple-tinged hairy bracts. The blue-lavender flowers have 3 petals, which are partially fused together. Fruit is an elliptical-shaped capsule containing a single seed.

Bloom season: Summer to early fall

Range/habitat: Widespread across North and South America and tropical Africa. In the United States, alligator flag is native to the Southeast in wet areas such as bogs, wetlands, ponds, marshes, and stream banks.

Comments: Thalia was the Greek goddess of comedy, but the genus honors Johannes Thal (1542–1583), a German physician and botanist. *Geniculate* ("jointed or bent like a knee") refers to the stems. These are host plants for the Brazilian skipper, and the common name refers to the plants growing where alligators live. The leaves contain rosmarinic acid, a compound that protects the leaves from ultraviolet light and is very similar to compounds found in some culinary herbs such as basil and oregano. The common name flag refers to the large light green leaves.

STEVE R. TURNER

AMERICAN BLUE HEARTS

Buchnera americana

Broomrape family (Orobanchaceae)

Description: Annual, hemiparasitic, 16"–36" tall; stem bears small hairs. The mainly unbranched stem and leaves have fine hairs. Lance-shaped leaves are up to 2½" long, have 3 main veins, and are serrated along the margins. The green to purplish 5-toothed calyx is also covered with fine hair. Tubular-shaped flowers have 5 flaring petals (petallike lobes) that are bluish purple (sometimes white) and have a hairy throat. Fruit is a dark purple capsule.

Bloom season: May bloom year-round; peak is midsummer to early fall.

Range/habitat: Eastern United States and Ontario, Canada, in moist sandy sites in prairies, open woodlands, and limestone glades

Comments: *Buchnera* honors Andreas Elias von Büchner (1701–1769), a German physician. *Americana* ("of America") refers to the plant's distribution. These blue hearts may or may not grow in association with a host plant, often an oak, beech, or cottonwood tree, and are partially parasitic to that host. Bees and butterflies are attracted to the flowers as pollinators.

NAKED BROOMRAPE

Orobanche uniflora

Broomrape family (Orobanchaceae)

Description: Parasitic plant, 1"–4" tall. Basal leaves are lance-shaped and up to ½" long. The long, yellowish flowering stalk has sticky hairs and bears a single flower. Tubular flowers are bluish or purple, 2" long, and have 2 prominent yellow ridges on the lower lip. Fruit is a capsule.

Bloom season: Early spring to midsummer

Range/habitat: Across much of North America in moist (seasonally) meadows, grasslands, open areas, and open woods

Comments: *Orobanche* is from the Greek *orobos* ("a type of vetch or climbing plant") and *ancho* ("to strangle"); both refers to the plant's parasitic nature. *Uniflora* refers to the single flower borne on the flowering stalk. This plant parasitizes species of stonecrop (*Sedum*), saxifrage (*Saxifraga*), and some Sunflower family members (Asteraceae). A dumbbell-shaped stigma (often yellow) on the upper inside roof of the flower resembles an anther and encourages insect pollinators to advance into the flower.

CRAIG MARTIN

PURPLE-WHITE OWL-CLOVER

Orthocarpus purpureoalbus
Broomrape/Paintbrush family (Orobanchaceae)

Description: Semiparasitic annual, 5"–20" tall. Stems may be branched halfway up the reddish-purple stems. Linear-shaped leaves are entire or cleft into 3 narrow segments and ¾"–1¾" long. Solitary, purple-white flowers, ¼"–½" long, arise from the leaf axils. The 2-lipped flowers have a purple-pink upper lip that is beaked and a broader lower lip that is white and pouch-like; the flowers become more purplish as they age. Leafy bracts are similar in shape to leaves and subtend the flowers. Fruit is a capsule.

Bloom season: Early spring to midsummer, depending on elevation

Range/habitat: Southwestern United States in meadows, moist grasslands, and open woodlands

Comments: *Orthocarpus* is from *ortho* ("straight") and *carpus* ("fruit"), for the seed capsule. *Purpureoalbus* ("purple white") is in reference to the flower colors. Originally placed in the Figwort family (Scrophulariaceae), these plants have been moved to the Broomrape family. This plant was first collected for science in New Mexico in the late 1860s.

WOODLAND BEARDTONGUE

Nothochelone nemorosa
Plantain family (Plantaginaceae)

Description: Perennial, 20"–40" tall from a woody base. Upright stems bear opposite, lance- to egg-shaped leaves that are serrated along the margins, short stalked, and 2"–5" long. Flowering stalks bear small clusters of blue-purple or pinkish-purple flowers about 1½" long, 2-lipped, and tubular. The lower lip projects downward and beyond the edge of the upper lip. Flowers have sticky hairs on the outside, and stamens are covered with hairs on the tip. Fruit is a capsule.

Bloom season: Early to midsummer

Range/habitat: Western United States in forests and rocky slopes

Comments: *Nothochelone* is from *notho* ("false") and *chelone* (the name of another genus) and refers to this plant being a "false" *Chelone*. *Chelone* ("turtle") refers to the flower resembling a turtle's head when viewed from the front; hence another common name: turtlehead. Formerly placed in the Figwort family (Scrophulariaceae).

STEVE R. TURNER

BLUE TOADFLAX

Nuttallanthus canadensis
Plantain family (Plantaginaceae)

Description: Annual or biennial, 10"–40" tall; stems and flowering stalks are green or reddish green. Linear-shaped leaves are up to 1½" long, entire, and stemless; upper leaves are alternate and lower leaves are opposite. Flowering stalks bear ½"-long, 2-lipped lavender flowers that have a slight nectar spur. Upper lip has 2 rounded lobes; the lower lip has 3 rounded and spreading lobes. Throat has 2 rounded ridges that are white. Fruit is a capsule with round, flat seeds.

Bloom season: Mid-spring to midsummer

Range/habitat: Eastern North America from Ontario, Canada, to Florida in sandy prairies, dunes, savannas, and fields

Comments: *Nuttallanthus* honors Thomas Nuttall (1786–1859), an English botanist, zoologist, and ornithologist who became a curator of the Harvard Botanic Gardens. Nuttall collected plants and animals on several western expeditions and in 1816 wrote *The Genera of North American Plants*. *Canadensis* ("of Canada") refers to the plant's distribution. Southern toadflax (*N. texensis*) is similar but has larger flowers and seeds with bumps. Blue toadflax is a larval host to the buckeye butterfly (*Junonia coernia*). Also known as Canadian toadflax.

TOM LEBSACK

COBAEA BEARDTONGUE

Penstemon cobaea
Plantain family (Plantaginaceae)

Description: Perennial, 1'–2½' tall and clump forming. Lance-shaped leaves are 1½"–6" long, covered in fine white hairs, serrated along the margins, and oppositely arranged along the stem. The upper leaves clasp the stem. Tubular-shaped flowers are white to pink to purple, 2" long, and borne on rigid, hairy stems. Flowers have 2 large lobes on top and 3 smaller lobes below; lobes may have white markings and purplish lines. Fruit is a black capsule.

Bloom season: Spring

Range/habitat: Central United States, mostly in prairies, meadows, rocky hillsides, and limestone outcrops

Comments: *Penstemon* is from *pen* ("almost") and *stemon* ("stamen"), which refers to the sterile stamen, called a staminode, typical of this genus. *Cobaea* honors Father Bernabé Cobo (1572–1659), a Spanish Jesuit and naturalist who resided in South America and Mexico for many years and resident of America for many years; he wrote about the Inca Empire and was an early investigator of quinine as a treatment for malaria. Also known as dew flower or foxglove penstemon for its foxglove-like flowers.

STEVE R. TURNER

WHIPPLE'S BEARDTONGUE

Penstemon whippleanus
Plantain family (Plantaginaceae)

Description: Perennial, up to 2' tall. Basal leaves are long stemmed, oval, and slightly toothed along the margins. Upper leaves are broadly lance-shaped, clasping the main stem, opposite, and up to 3" long. Tubular flowers are white to a reddish dark purple and borne in ringlike clusters. The 5-lobed flowers are covered with rows of white hairs, as are the leafy bracts that subtend the flowers. Flower's 2 upper lobes curve slightly upward; the 3 lower lobes project forward and bear long white hairs on the inside.

Bloom season: Summer to early fall

Range/habitat: Rocky Mountain states from Montana to New Mexico in subalpine slopes and woodlands

Comments: *Penstemon* is from *pen* ("almost") and *stemon* ("stamen"), which refers to the sterile stamen, called a staminode, typical of this genus. *Whippleanus* honors Lt. Amiel Whipple (1816–1863), a member of the US Army Corps of Topographical Engineers who led an expedition along the 35th parallel to survey for a transcontinental railroad route from Oklahoma to California.

STEVE R. TURNER

AMERICAN SPEEDWELL

Veronica americana
Plantain family (Plantaginaceae)

Description: Perennial with smooth, round stems 4"–40" long. Lance-shaped leaves are ½"–3¼" long, opposite, and toothed or entire along the margins. Small saucer-shaped flowers, ⅓" wide, are blue, violet, or white, with 4 petals (the upper petal is 2 petals fused together) and 2 stamens. Fruit is a 2-lobed capsule containing numerous seeds.

Bloom season: Late spring to late summer

Range/habitat: Widely distributed across North America (except for the southeastern United States) in streams, brooks, ponds, seeps, and riparian areas.

Comments: *Veronica* honors Saint Veronica, who provided Jesus with a cloth to wipe his face on his way to Calvary. *Americana* ("of America") refers to the plant's distribution. The common name speedwell refers to the rapid development and short life span of the flowers. The plant is edible and eaten like watercress. Also known as American brooklime for the plant's preference of growing in muddy edges of brooks.

MARGARET MARTIN

BLUE-HEAD GILIA

Gilia capitata
Phlox family (Polemoniaceae)

Description: Annual, arising on thin stems, 1'–3' tall. Basal and stem leaves are divided into narrow linear segments. Lower, larger leaves may be 5" long. The ¼"-wide blue flowers are arranged in rounded clusters about 1"–1½" wide. The 5 petals form a short tube that flares outward. Fruit is a somewhat rounded capsule.

Bloom season: Late spring to summer

Range/habitat: Western United States and parts of the upper midwestern and northeastern United States in open fields, meadows, rocky outcrops, and disturbed sites

Comments: *Gilia* honors Filippo Luigi Gilii (1756–1821), an Italian naturalist and director of the Vatican Observatory who coauthored books about South American plants. *Capitata* ("dense head") refers to the clustered arrangement of the flowers. Many types of pollinators, including bees, butterflies, wasps, and flies, are attracted to the flowers. Also known as bluefield gilia.

BIRD'S EYE GILIA

Gilia tricolor
Phlox family (Polemoniaceae)

Description: Annual, 4"–18" tall. Leaf segments are narrow and linear. Flowers have 5 green sepals and 5 fused petals. Blue-violet bell-shaped flowers are about ½" wide and have a yellow throat and dark purple ring that surrounds the throat opening. The 5 stamens have blue anthers, which also add to the flower's color. Fruit is a capsule.

Bloom season: Early to late spring

Range/habitat: California and Texas, although grown commercially for gardeners, which has expanded its range. Grows in dry to wet meadows, grasslands, mesa tops, and open slopes.

Comments: *Gilia* honors Filippo Luigi Gilii (1756–1821), an Italian naturalist and director of the Vatican Observatory who coauthored books about South American plants. *Tricolor* ("three colors") refers to the different colors on the flowers. The chocolate-scented flowers attract a number of pollinators, including butterflies, hummingbirds, and bees.

ALISON P. NORTHUP

NEEDLE-LEAF GILIA

Giliastrum acerosum
Phlox family (Polemoniaceae)

Description: Perennial, 3"–10" tall. Leaves are alternate, pinnate or pinnately cleft; the 2–7 segments are needlelike to linear and up to ½" long; upper leaves have fewer segments and are more palmately divided. The sharp-pointed leaflets have some sticky hairs. Small flowers are bowl-shaped, about ½" wide, and 5-lobed. The lobes are blue to purplish, with a white band near the yellow to greenish center. Yellow, heart-shaped stamens project beyond the flower's mouth and are quite noticeable. Fruit is an egg-shaped capsule.

Bloom season: Mid-spring through summer

Range/habitat: Southern United States from Arizona to Oklahoma in dry and rocky canyons, prairies, and hillsides

Comments: *Giliastrum* honors Filippo Luigi Gilii (1756–1821), an Italian naturalist and director of the Vatican Observatory who coauthored books about South American plants. *Acerosum* ("sharp" or "with stiff needles") refers to the leaves. Also known as bluebowls, for the shape of the flowers. Native Americans used the crushed leaves to treat cramped muscles.

STEVE R. TURNER

PALE-FLOWER SKYROCKET

Ipomopsis longiflora
Phlox family (Polemoniaceae)

Description: Annual or biennial, 6"–24" tall with smooth or slightly hairy stems. Basal leaves have narrow segments; upper leaves are narrow but with fewer segments. Lower leaves wither by the time the flowers mature. White to pale blue flowers are 1"–2" long and have a long, trumpetlike corolla tube that flares open to 5 petals. Stamens do not extend beyond the mouth of the corolla. Fruit is a capsule.

Bloom season: Early spring to mid-fall

Range/habitat: Southwestern United States from Arizona to Texas and north to South Dakota in open fields, sagebrush shrublands, dry plains, and woodlands up to 7,000' in elevation

Comments: *Ipomopsis* ("like-*Ipomoea*") refers to this skyrocket's flowers resembling those of flowers in *Ipomoea*, the genus for morning glory. *Longiflora* ("long flower") refers to the length of these flowers.

BORREGOWILDFLOWERS.ORG

BRISTLY GILIA

Langloisia setosissima
Phlox family (Polemoniaceae)

Description: Annual, 2"–10" tall. Linear to inversely lance-shaped leaves are spirally arranged around the stem, 1"–1½" long, densely hairy, with a serrated or lobed margin. The small, funnel-shaped flowers are white to pale blue in color, 5-lobed, and about 1" wide. Fruit is a small capsule.

Bloom season: Mid-spring to early summer

Range/habitat: Western United States in desert scrub, rocky areas, and washes

Comments: *Langloisia* honors the Reverend Father Auguste Barthélemy Langlois (1832–1900), a Louisiana priest and botanist. He described and collected lichens in central and southern Louisiana. *Setosissima* ("very bristly hairy") is for the hairy leaves. This species is the only one of the genus in the United States. Seeds become gelatinous when wet.

KATIE BYERLY

WILD BLUE PHLOX

Phlox divaricata
Phlox family (Polemoniaceae)

Description: Perennial, 12"–15" tall with hairy and sticky stems. Lance- to elliptical-shaped opposite leaves are up to 2" long and hairy. Tubular-shaped flowers flare open with 5 petallike notched lobes. Flowers are blue, lavender, rose, or white and about 1½" wide. Fruit is a capsule.

Bloom season: Spring and early summer

Range/habitat: Eastern North America in woods, fields, and along stream banks

Comments: *Phlox* ("flame") refers to the brilliant colors of certain species in the genus. *Divaricata* ("spreading") refers to the plant's habit of spreading by lateral shoots growing along the ground and rooting at the nodes. Butterflies and long-tongued bees are the primary pollinators, since nectar rewards are found deep within the corolla tube. Also known as woodland phlox or wild sweet William.

STEVE R. TURNER

LONG-LEAVED PHLOX

Phlox longifolia
Phlox family (Polemoniaceae)

Description: Perennial, 4"–20" tall. Narrow linear leaves are up to 3" long, pointed at the tip, and may have a few hairs. Flowering stalks are covered with sticky glands and bear trumpet-shaped flowers with a long tube and 5 flaring lobes that may be notched or not. Calyx has a raised white keel. White to lavender to pink flowers are 1"–1½" wide and have a white inner circle. Fruit is a capsule.

Bloom season: Mid-spring to midsummer

Range/habitat: Western North America in dry foothills and sagebrush habitats

Comments: *Phlox* ("flame") refers to the brilliant colors of certain species in the genus. *Longifolia* ("long leaved") refers to the leaves. Nathaniel Wyeth (1802–1856), known as the Cambridge Iceman due to his Boston ice business, first collected this plant in 1832 on his return journey from the Oregon Territory to Massachusetts. Rocky Mountain phlox (*P. multiflora*) is similar but grows at higher elevations and lacks the white keel on the calyx.

SPREADING JACOB'S LADDER

Polemonium reptans
Phlox family (Polemoniaceae)

Description: Perennial, 12"–18" tall. Alternate compound leaves bear 5–15 oval to egg-shaped leaflets about 1" long. Bell-shaped flowers range from blue to lavender to pink to white in color and are borne in loose clusters along weak stems. Flowers are ¾"–1" wide with 5 rounded petals. Fruit is a capsule.

Bloom season: Mid-spring to early summer

Range/habitat: Midwestern and eastern United States and Canada in deciduous woodlands, often along waterways

Comments: *Polemonium* may be named for Polemon, a Greek philosopher and healer; a species of this genus was associated with him due to its medicinal properties. *Reptans* ("creeping") refers to the growth habit of this plant. The common name refers to the leaflets being arranged like the rungs of a ladder the biblical Jacob dreamt about that the angels used to ascend to and descend from heaven. The roots have been used medicinally to treat kidney disease and as a diuretic. Western Jacob's ladder (*P. occidentale*) is similar and grows across most of western North America.

GARY PAULL

SKY PILOT

Polemonium viscosum
Phlox family (Polemoniaceae)

Description: Perennial, 5"–15" tall. Stems and leaves are covered with sticky hairs and have a pungent odor. Compound fernlike leaves that point upward have numerous spoon-shaped leaflets. Flowering stems, 4"–20" tall, arise from the leaves and bear fanlike clusters of blue or occasionally white funnel-shaped flowers with a short floral tube. Stamens with golden-orange anthers project beyond the flower opening. Fruit is a capsule.

Bloom season: Spring and early summer

Range/habitat: Western North America at high elevations in rocky alpine sites

Comments: *Polemonium* may be named for Polemon, a Greek philosopher and healer; a species of this genus was associated with him due to its medicinal properties. *Viscosum* ("sticky leaves") refers to the texture of the leaves. Thomas Nuttall (1786–1859), a British botanist and explorer, first collected this plant for science in the headwaters of the Platte River during his 1834–1837 western explorations. Sweet and/or odoriferous flowers may be present. The common name refers to the high-elevation distribution of these plants. Several other species of *Polemonium* grow at high elevations, including dense sky pilot (*P. confertum*) in the Rocky Mountains.

DAVID LEGROS

PICKERELWEED

Pontederia cordata
Pickerelweed family (Pontederiaceae)

Description: Perennial, emergent aquatic plant, 2'–4' tall. Highly variable arrowhead-shaped leaves are up to 10" long and 5" wide and have heart-shaped bases and smooth edges. Bluish tubular-shaped flowers are borne in short clusters on stout, hollow flowering stalks that rise 1'–3' above the water's surface; stalks bear a single leaf. Flowers are ½" wide, have 6 spreading lobes; the upper lobe has yellowish patches that act as nectar guides. Flowering spikes, up to 5" long, droop after maturity and release the large seeds with toothed edges into the water.

Bloom season: Summer to fall

Range/habitat: Eastern North America and the Caribbean in shallow aquatic habitats such as ponds, swamps, sloughs, and bogs

Comments: *Pontederia* is for Giulio Pontedera (1688–1757), a professor of botany at the University of Padua, the same location where Galileo Galilee lectured from 1592 to 1610. *Cordata* ("heart-shaped") refers to the leaf bases. The seeds and leaves are edible.

WESTERN MONKSHOOD

Aconitum columbianum
Buttercup family (Ranunculaceae)

Description: Perennial, 3'–7' tall. Maplelike leaves, 2"–7" long, are divided into 3–7 lobes that may be toothed or lobed again. Dark blue flowers, 1"–2" long, are borne at the tips of the flowering stalk and have 5 petallike sepals and 2 petals. The upper sepal is hoodlike, and the petals form a spur. Fruit is a capsule.

Bloom season: Mid- to late summer

Range/habitat: Western North America in moist meadows, stream banks, and woodlands.

Comments: *Aconitum* was an ancient Greek word used for this plant. *Columbianum* ("of western North America") refers to the wide distribution. The plant is toxic and might have been used to treat arrows or to poison wolves, resulting in another common name: wolf's bane. The hoodlike sepals resembled the hood on a monk's cloak; hence the common name.

C. PIEFER

CAROLINA ANEMONE

Anemone caroliniana
Buttercup family (Ranunculaceae)

Description: Perennial, 3"–12" tall. Palmately compound leaves have 3 leaflets that are notched at the tip or deeply lobed, slightly hairy (or hairless), 1" wide, and with a 1"- to 4"-long stem. Leaflets may also be deeply notched or lobed again. The basal leaves have a whorl of 3 leafy bracts along the lower portion of the stem. The flowering stalk bears a single 1"- to 1½"-wide flower that has 8–20 petallike sepals of varying lengths and may be blue to white to lavender in color. Back side of the sepals and upper portion of the flowering stem have long hairs. Center of the flower is greenish and has numerous stamens. Fruit is a woolly seed.

Bloom season: Spring

Range/habitat: Central and southeastern United States in dry prairies, open woodlands, and barrens

Comments: *Anemone* honors the Anemoi, Greek gods of the four winds. Many species in this genus are referred to as "windflowers" for the wind-borne dispersal of the seeds. *Caroliniana* refers to the plant first being collected for science in the Carolinas region. The highly variable flowers may be two-toned, blue and white, as well.

STEVE R. TURNER

ROCKY MOUNTAIN COLUMBINE

Aquilegia coerulea
Buttercup family (Ranunculaceae)

Description: Perennial, 8"–24" tall. Fernlike leaves are divided and lobed. The 3"-wide flowers are bicolored, with the sepals differing from the petals; flowers are variable in color, from light blue to white to pale yellow. Flowers have 5 sepals and 5 petals, which form long spurs, and 50–130 stamens. Fruit is a capsule.

Bloom season: Spring

Range/habitat: Western North America in the Rocky Mountain states in meadows and woodlands

Comments: *Aquilegia* ("eagle") refers to the floral spurs resembling an eagle's talons. *Coerulea* ("sky blue") refers to the flower color. This is the state flower of Colorado, also known as the variant *A. caerulea.* Hawkmoths, butterflies, and long-tongued bees pollinate these flowers.

STEVE R. TURNER

FAIRY HATS

Clematis crispa
Buttercup family (Ranunculaceae)

Description: Vine, 6'–10' tall. Leaves are opposite and pinnately compound and have 3–5 linear to egg-shaped leaflets that are 1¼"–4" long. The lavender to bluish-purple, bell-shaped flowers are 1"–2" long and hang downward. Flowers lack petals and instead have sepals that are joined together to form an urn shape; each wavy or frilled lobe curls backward and to the side. Fruit is a feathery seed.

Bloom season: Spring, but may have second bloom in fall and occasionally in summer

Range/habitat: Southeastern United States in swamps, flooded woodlands, and marshes

Comments: *Clematis* is from the old Greek word *klematis* ("vines"). *Crispa* ("curled") refers to the petallike sepals. The tendrils make this plant a good climber. Also known as swamp leather-flower or blue jasmine.

MARGARET MARTIN

SIERRA LARKSPUR

Delphinium glaucum
Buttercup family (Ranunculaceae)

Description: Perennial with stout stems that may reach 9' tall. Basal leaves are palmately lobed, often into 5 divisions, with each lobe toothed to varying degrees along the margin. Flowers are borne along an elongated stalk, from a few to around 50, and are dark blue to purplish. The 5 sepals are angled outward at the tip; the spur formed by the upper sepal is about ½" long. Four petals are smaller; the upper 2 also form a spur, which is concealed by the sepals. Fruit is a 3-chambered pod.

Bloom season: Summer

Range/habitat: Western North America in moist montane areas and woodlands

Comments: *Delphinium* is from the Latin *delphinus* ("dolphin"), referring to the shape of the flower buds. *Glaucum* ("smooth") refers to the stems. These plants contain toxic alkaloids. Also known as tall or mountain larkspur. The common name larkspur refers to the backward-projecting floral spur, which resembles the rear "spur" on a lark's foot.

PATRICK ALEXANDER

NUTTALL'S LARKSPUR

Delphinium nuttallianum
Buttercup family (Ranunculaceae)

Description: Perennial, up to 20" tall. Basal leaves are highly divided and lobed; often wither before the flowers open. Flowering stalks are 6"–12" long and bear light blue to dark purple flowers with a backward-projecting spur. Fruit is a 3-chambered pod.

Bloom season: Spring to early summer

Range/habitat: Western North America in grasslands, sagebrush shrublands, meadows, and woodlands

Comments: *Delphinium* is from the Latin *delphinus* ("dolphin") and refers to the shape of the flower buds. *Nuttallianum* honors Thomas Nuttall (1786–1859), an English botanist and explorer who collected and studied plants across the United States from 1807 to 1841 before retiring back to England. Like many larkspurs, Nuttall's larkspur contains toxic alkaloids. The blue flowers were used in making a blue dye.

JACOB W. FRANK, NPS

AMERICAN PASQUEFLOWER

Pulsatilla patens

Buttercup family (Ranunculaceae)

Description: Perennial, up to 1' tall. Fernlike foliage is covered in soft white hairs. Flowering stalks elongate after the flower buds develop. Bell-shaped flowers are generally bluish purple, with 6–8 petallike sepals and numerous stamens. Fruit is a seed with a long, feathery plume.

Bloom season: Spring

Range/habitat: Much of western and northern North America in prairies, meadows, and rocky areas

Comments: *Pulsatilla* ("flowers which sway in the wind") describes the flowers. *Patens* ("spreading") refers to the sepals. American pasqueflower is North Dakota's state flower and the provincial flower of Manitoba, Canada. The plant was and still is used medicinally. *Pasque* is an old French word for Easter, referring to the blooming period of these flowers. Also known as Eastertime pasqueflower.

STEVE R. TURNER

BLUETS

Houstonia caerulea

Madder family (Rubiaceae)

Description: Perennial, low-growing tufts, 4"–8" tall. The smooth-margined leaves are small, about ½" long, spatula-shaped to linear, and form a basal rosette. The unbranched flowering stalk bears pairs of ¼"-long leaves, which are linear and opposite. The flowering stalk usually bears 1 flower, but there may be 2 per stem. The small, pale bluish or sometimes white flowers have 4 petallike lobes with yellow centers, 4 tube-shaped sepals, and are about ½" wide. Fruit is a 2-lobed capsule with minute bumps.

Bloom season: Mid-spring to midsummer

Range/habitat: Eastern North America in sandy areas in prairies, savannas, woodlands, and along streams

Comments: *Houstonia* honors William Houstoun (1695–1733), a Scottish botanist and surgeon who collected plants in Mexico, the West Indies, and South America. *Caerulea* ("sky blue") refers to the color of the flowers. These plants produce two types of flowers: those with long stamens and a short style and others with short stamens and a long style. For both types of flowers, the stamens and styles generally do not project much, if at all, above the flower center. Also known as Quaker ladies for the resemblance of the flowers to hats worn by Quaker women.

TOM LEBSACK

PURPLE GROUND-CHERRY

Quincula lobata

Nightshade family (Solanaceae)

Description: Perennial, with ridged stems that spread 1'–2' long. The egg- to lance-shaped leaves are 1"–4" long, often with notched or ruffled edges with rounded lobes. Flowers are bell- to flat-shaped, ¾"–1½" wide, purple-blue, and often with white throats. Along with the 5 petals, there are 5 stamens with yellow anthers that stick up above the flower. Fruit is an inflated lantern-like structure that contains a greenish-yellow, cherry-size berry.

Bloom season: Spring to fall until first frost

Range/habitat: Southwestern United States and northern Mexico in dry sites in prairies, fields, grasslands, woodland edges, and open forests

Comments: *Quincula* ("five spots") refers to the white spots on the petals. *Lobata* ("having lobes") refers to the edges of the leaves. Indigenous people harvested the edible seeds. Also known as Chinese lantern for the papery pods that surround and protect the berry.

KATIE BYERLY

BLUE VERVAIN

Verbena hastata

Verbena family (Verbenaceae)

Description: Biennial or perennial, 2'–5' tall with square, hairy stems. Lance-shaped leaves are up to 6" long, opposite, and sharply toothed along the margins. Flowering stems branch like a candelabra, bearing narrow 2"- to 6"-long pencillike spikes of flowers. The ⅛"- to ¼"-long blue, 5-lobed, tubular-shaped flowers are densely packed along the spike. Fruit is a nutlet.

Bloom season: Summer

Range/habitat: Eastern North America in wet meadows, stream banks, fields, and sloughs

Comments: *Verbena* ("sacred plant") refers to the medicinal use of members of this genus to cure all sorts of maladies. *Hastata* ("spear shaped") refers to the leaf shape. Native Americans harvested the nutlets for food and the leaves for teas or eating.

JIM FOWLER

SAGEBRUSH VIOLET

Viola beckwithii

Violet family (Violaceae)

Description: Perennial, 2"–20" tall; much of the stem may be underground. Leaves are fleshy, compound with narrow leaflets, and borne on long stems. The 1"- to 1½"-long flowers are bicolored and show a fair amount of variation. The upper 2 petals are maroon to purple; the lower 3 are lavender to bluish (sometimes white) with yellow centers and purple nectar lines. Lowest petal is spurred. Fruit is a seed.

Bloom season: Early spring to midsummer

Range/habitat: Western United States in sagebrush steppe and pine woodlands

Comments: *Viola* is the Latin name for several different scented flowers. *Beckwithii* is for Lt. Edward Beckwith (1818–1881), who served with the US Army Corps of Topographical Engineers and led an expedition along the 38th and 39th parallels searching for a potential railroad route to California. One of the early spring wildflowers in the sagebrush steppe, this plant is also known as the Great Basin violet.

BIRD-FOOT VIOLET

Viola pedata

Violet family (Violaceae)

Description: Perennial, 4"–10" tall and often growing in clumps. The deeply divided leaves are borne on stems 4"–6" long, are ¾"–2" long and palmately lobed with 3–5 segments. Each segment is lobed again 2–3 times. The upper end of the lobes is toothed and often wider than the base. The ¾"- to 1½"-long flowers are bicolored and variable, with the upper 2 lobes generally dark purple and the lower 3 lobes bluish with white around the flower's throat. Fruit is a seed.

Bloom season: Spring to early summer

Range/habitat: Central and eastern North America in rocky areas in woods, roadsides, and disturbed areas

Comments: *Viola* is the Latin name for several different scented flowers. *Pedata* ("like a foot") is for the resemblance of the leaves to a bird's foot. This violet reproduces solely by seed, not vegetatively like many other *Viola* species. The seed has a gelatinous cover that attracts ants, which haul the seeds back into their underground nests as a form of dispersal.

CRAIG MARTIN

PRAIRIE VIOLET

Viola pedatifida
Violet family (Violaceae)

Description: Perennial, 3"–12" tall. Basal leaves, 2–12, are about 1" wide and deeply palmately lobed with linear lobes. The leafless flowering stem rises above the leaves, and the stalk curves downward where the flower bud forms. The bluish-purple, ¾"-wide flowers have 5 petals and 5 green, pointed sepals. The 2 upper petals are long and rounded, the lower 3 are narrower, and the middle petal has purplish nectar guide lines. The flower's throat has fine white hairs. Fruit is a capsule.

Bloom season: Spring and possibly again in the fall

Range/habitat: South-central Canada and the central United States in prairies and savannas

Comments: *Viola* is the Latin name for several different scented flowers. *Pedatifida* ("palmately divided into cleft segments") refers to the leaflets, which resemble a bird's foot with the toes divided again. Also known as crow-foot violet or larkspur violet for the similarity of the leaflets to the divided leaves of members of the *Delphinium* genus. Similar to many other violets, prairie violet also has self-fertilizing (cleistogamous) flowers that form triangular seedpods, which eject the seeds when mature.

JIM FOWLER

COMMON BLUE VIOLET

Viola sororia
Violet family (Violaceae)

Description: Perennial, 6"–10" tall and often growing in clumps. Glossy leaves are heart-shaped with wavy edges, 3" wide, and form a basal rosette. Bluish-purple flowers (some are white) arise on flowering stalks that barely rise above the leaves. Flowers have 5 petals, which are hairy to various degrees. The upper 2 petals spread outward, the 2 lateral petals have thick white hairs near the flower's throat, and the central petal has hairs and a white spot with purplish nectar lines. Fruit is a capsule.

Bloom season: Spring

Range/habitat: Eastern North America in moist woodlands, savannas, or swamps

Comments: *Viola* is the Latin name for several different scented flowers. *Sororia* ("sisterly") refers to this plant resembling many other violets. Both the leaves and flowers are edible. The flowers have been made into sweets; the roots have been used to treat coughs or respiratory illnesses. Like many other violet species, common blue violet also forms self-fertilizing (cleistogamous) flowers that form later in the spring and summer. The hairs on the petals serve to restrict raindrops from flowing into the flower and diluting the nectar, and as a grip for insects visiting the flowers. Common blue violet, also known as woolly blue violet or wood violet, is the state flower of Wisconsin, New Jersey, Illinois, and Rhode Island. Another color form called Confederate violet resembles the white-gray uniforms of Confederate soldiers.

GREEN AND BROWN FLOWERS

ELEANOR DIETRICH

This section includes flowers that are predominately green or brown. Some flowers in this section also tend toward yellow, lavender, or pale purple. Check those sections if you do not find the flower you are looking for in this section.

BONNIE ISAAC

AMERICAN SWEETFLAG

Acorus americanus
Calamus family (Acoraceae)

Description: Perennial, emergent growth, 1'–6' tall. The narrow swordlike leaves are up to 4¾' long, bright green, with 2–6 prominent veins, and, when viewed in cross section, have a swollen center. Leaves arise from a white base that may be tinged with red. Minute yellowish-green flowers are arranged along a fleshy stalk (called a spadix) that is 2"–5" long. The flowers have 6 sepals and tepals. Lacking a spathe, the plant has a specially shaped leaf that partially surrounds the spadix (called a sympodial leaf). Pollen stains an aniline blue color. Fruit is a brown to reddish-colored berry with a gelatinous inside bearing 6 seeds.

Bloom season: Mid-spring to midsummer

Range/habitat: Across southern Canada and the northern United States in swamps, ponds, wetlands, and stream banks

Comments: *Acorus* has its origins in the Greek *coreon* ("pupil"), which refers to the plant's use as a medicinal treatment for eye inflammation. *Americanus* ("of America") refers to the distribution of this plant in North America. Native Americans used this plant for medicine, in ceremonies, and as a trade item. Its extensive distribution in the United States and Canada may be due to Indigenous peoples. Calamus oil is derived from the roots and used as a tonic for digestive issues. The citrus-like aroma of the leaves has made its way into the beer making industry.

TOM HARRINGTON

FILMY ANGELICA

Angelica triquinata
Carrot family (Apiaceae)

Description: Perennial, up to 6' tall; stems become reddish with age. Leaves are pinnately or bipinnately compound; leaflets are toothed along the margins and with a pointed tip. Each leaflet may be up to 3" long and have hairy margins. The base of the leaf stem has a sheath that encompasses the main stem and appears "filmy." Small clusters, 13–25, of tiny greenish-white flowers (umbellets) compose a larger umbel. Fruits are striped seeds with winged margins.

Bloom season: Late summer to early fall

Range/habitat: Southeastern United States in mountain meadows and woodlands

Comments: *Angelica* ("angel") refers to the medicinal and angelic protection against evil properties as revealed to humans by an archangel. *Triquinata* refers to the 5-parted triangular leaves. Insects such as bees and flies that obtain nectar from the flowers often appear lethargic or drugged, and people may develop an allergic reaction or dermatitis upon contact with the plants. Also known as mountain angelica for its high-elevation distribution.

NATHAN RAUH

CANADIAN BLACK-SNAKEROOT

Sanicula canadensis

Carrot family (Apiaceae)

Description: Biennial or perennial, 1'–2' tall. Leaves are alternate, rounded in outline, 3"–5" wide; lower leaves are palmately divided into 3 segments. Lateral leaflets may be deeply divided into 2 lobes that are broadly elliptical to egg-shaped in outline and double toothed along the margins. Upper compound leaves have shorter stems and are narrower. Flat-topped clusters of greenish-white flowers are divided into 1–4 smaller heads; each cluster is about 2½" wide and irregularly round. Flowers are either male or perfect (both male and female), which have ovaries with hooked bristles. Fruit is a bur-like pod.

Bloom season: Late spring to midsummer

Range/habitat: Central and eastern North America in shady locations in moist woodlands, field edges, thickets, and north-facing slopes

Comments: *Sanicula* ("to heal") refers to the reported medicinal use of some species in this genus. *Canadensis* ("of Canada") refers to the plant's distribution. Several similar-looking species of *Sanicula* occur in North America and are distinguished from one another by minor features.

KATHY RIGALL

ANTELOPE HORNS

Asclepias asperula

Dogbane family (Apocynaceae)

Description: Perennial, arising from a stout root. Stems may grow upright or trail along the ground. Plant is 10"–20" tall and bears lance-shaped leaves, 5"–10" long; leaves are pointed at the tip and generally folded lengthwise. Rounded flower clusters have greenish-white petals, sometimes tinged with purple. Lobes are about ½" long. Greenish to purplish hoods are club- to sickle-shaped and abruptly curved from the anthers. Fruit is a slender pod, 2"–6" long.

Bloom season: Spring to early summer

Range/habitat: Western to central United States in sandy sites or rocky areas in sagebrush, woodlands, and mountain forests

Comments: *Asclepias* refers to Asklepios, a mortal physician who was an authority on plants and their healing properties. According to Greek mythology, Zeus killed Asklepios with a thunderbolt after the physician boasted about being able to revive the dead. *Asperula* ("rough") refers to the texture of the leaves. The unique flowers are pollinated by a variety of insects. As the seedpods mature, they start to curve and resemble a pronghorn's (or "antelope's") horns; hence the common name.

PALLID MILKWEED

Asclepias cryptoceras

Dogbane family (Apocynaceae)

Description: Perennial, stems upright or trail along the ground, 4"–12" long. Leaves opposite, broadly oval, fleshy, 1"–4½" long and nearly as wide, and smooth. Clusters of greenish-white flowers with 5 linear sepals and 5 reflexed greenish-white petals arise toward the top of the stem. Corona consists of 5 stamens encased in rose-red hoods. Spindle-shaped pods are up to 2½" long.

Bloom season: Mid-spring to early summer

Range/habitat: Western United States in dry semi-shrub, sandy, and rocky areas

Comments: *Asclepias* refers to Asklepios, a mortal physician who was an authority on plants and their healing properties. According to Greek mythology, Zeus killed Asklepios with a thunderbolt after the physician boasted about being able to revive the dead. *Cryptoceras* ("hidden horn") refers to the hornlike projection hidden in the hood. The legs of small pollinators may become stuck in the hood slits, resulting in the insect's demise. Large pollinators such as butterflies, bees, and larger moths can withdraw their legs, which can have sacs of pollen attached to them. Stems bleed a milky latex when broken or crushed.

ELEANOR DIETRICH

VELVETLEAF MILKWEED

Asclepias tomentosa

Dogbane family (Apocynaceae)

Description: Perennial, 2'–3' tall with stiff stems. Elliptical to oval-shaped leaves are opposite, covered with soft hairs, 2"–4" long, with pink midveins. The greenish-white flowers are borne in a dense head; each flower is about ½" long. Fruit is an elongated pod.

Bloom season: Mid-spring to late summer

Range/habitat: Southern United States from Texas to Florida in moist woods and sandhills

Comments: *Asclepias* refers to Asklepios, a mortal physician who was an authority on plants and their healing properties. According to Greek mythology, Zeus killed Asklepios with a thunderbolt after the physician boasted about being able to revive the dead. *Tomentosa* ("densely covered with short hairs") refers to the leaves, which are covered with soft hairs. Though not as showy as some of the other milkweeds, this species also attracts butterflies (monarchs included) and a host of other insects to their flowers. Also called tuba milkweed for the shape of the flower hoods resembling a tuba.

ELEANOR DIETRICH

GREEN COMET MILKWEED

Asclepias viridiflora

Dogbane family (Apocynaceae)

Description: Perennial, up to 25" tall with unbranched stems. Lance-shaped leaves are up to 5" long, folded lengthwise along the midrib, light green below, and with or without hairs. Leaves also have a wavy pattern. From the leaf axils arise a 2"-wide, dome-shaped cluster of 15–80 greenish flowers that become purple tinged as they mature. The ⅓"-long flowers have 5 reflexed sepals and petals and 5 upright hoods that surround a central column. Fruit is a smooth slender pod, 3½"–5" long.

Bloom season: Late spring to midsummer

Range/habitat: Eastern and central United States and southern Canada in prairies, savannas, roadsides, and open woodlands

Comments: *Asclepias* refers to Asklepios, a mortal physician who was an authority on plants and their healing properties. According to Greek mythology, Zeus killed Asklepios with a thunderbolt after the physician boasted about being able to revive the dead. *Viridiflora* ("green flowers") describes the color of the flowers. The common name comes from the flower head resembling a comet's head and tail. Broken stems bleed a milky latex.

CHUCK TAGUE

GREEN DRAGONS

Arisaema dracontium

Arum family (Araceae)

Description: Perennial, 1'–4' tall. Produces 1 leaf, but the leaf stem, which may reach 20" long, may fork to resemble another leaf. Leaf is divided into 5–15 leaflets of different lengths, which are palmately arranged. The flowering stalk bears a 6"-long spadix that extends beyond the flowers and green leaflike spathe. Tiny greenish-yellow flowers crowd the lower portion of the spadix. Fruit is a reddish berry.

Bloom season: Spring

Range/habitat: Central and eastern United States and south-central Canada in boggy areas, along creek beds, and in moist locations in woods

Comments: *Arisaema* is from the Greek *aris* ("arum") and *aima* ("red spotted"), referring to the *Arum*-like leaves of certain species that also have red blotches. *Dracontium* ("small dragon") refers to the compound leaflets, which resemble dragon's wings in profile. The plant is toxic, containing calcium oxalate crystals that can cause burning of the mouth if ingested. Green dragon flowers on a particular plant may be of one sex or both sexes, and the plants may change that sex year after year; this is known as "gender diphasy."

STEVE R. TURNER

JACK-IN-THE-PULPIT

Arisaema triphyllum
Arum family (Araceae)

Description: Perennial, 1'–2' tall. Two large, 1'- to 1½'-long leaves arise from a single stalk and form an umbrella over the flowers. Divided into 3 segments, leaflets are 3"–5" long. The spadix ("Jack") has numerous tiny green to purplish flowers along an extended spike that is surrounded by a green or purplish sheathlike spathe ("the pulpit"), which is open at the bottom to expose the flowers and forms a hood over the top of the spadix. The inside portion of the spathe is streaked with purple and greenish-white lines. Fruit is a reddish berry borne in a cluster.

Bloom season: Spring

Range/habitat: Eastern North America in moist thickets and woodlands.

Comments: *Arisaema* is from the Greek *aris* ("arum") and *aima* ("red spotted"), referring to the *Arum*-like leaves of certain species that also have red blotches. *Triphyllum* ("three leaved") refers to the leaves. Initially, the flower produces male flowers, which become hermaphroditic, with female flowers below and male flowers above on the spadix. The roots contain toxic calcium oxalate compounds, which cause burning of the mouth and throat.

ELK CLOVER

Aralia californica
Ginseng family (Araliaceae)

Description: Perennial, 6'–9' tall on thick, nonwoody stems. Huge leaves are pinnately or tripinnately compound and 3'–6' long. Leaflets are oppositely arranged except for the terminal leaflet, 6"–12" long and toothed along the margins. Small greenish-white flowers are borne in an overall umbrellalike cluster, with each smaller cluster appearing round or starlike. Individual flowers are ⅛" wide. Fruit is a ⅛"-long purple to black berry containing 3–5 seeds.

Bloom season: Midsummer to early fall

Range/habitat: Oregon and California in moist forests, riparian areas, seeps, and chaparral

Comments: *Aralia* is a Latinized name derived from an old French-Canadian word, *aralie. Californica* ("of California") refers to the plant's distribution and type locality. Medicinally harvested for use as a cough suppressant, arthritis treatment, and as an anti-inflammatory. Also called California spikenard.

STEVE R. TURNER

WILD SARSAPARILLA

Aralia nudicaulis
Ginseng family (Araliaceae)

Description: Perennial, 1'–2' tall with smooth stems. Large leaves arise from the base on a long stem and are pinnately compound. Usually twice divided, the leaves are first divided into 3s and then again into 5 (3–7) egg-shaped leaflets that are 2"–5" long. Leaflets are finely toothed along the margins and pointed tips. Tiny ⅛"-wide greenish-white flowers are borne in rounded clusters (umbels), 1½"–2" wide. The clusters contain 20–40 flowers, which have 5 recurved petals and 5 exserted stamens. Fruit is a purplish-black berry when mature.

Bloom season: Mid-spring to early summer

Range/habitat: Northern and eastern North America along bog edges, thickets, and woodlands

Comments: *Aralia* is a Latinized name derived from an old French-Canadian word, *aralie. Nudicaulis* is from two Latin words meaning "naked stalk," referring to the smooth, hairless stems. The roots have been used in herbal medicine and sodas as a substitute for true sarsaparilla (*Smilex* sp.). The edible berries are a little spicy and sweet. The leaves change color with the seasons, first appearing bronze, then green in summer, and red or yellow in fall. American spikenard (*A. racemosa*) grows in much of the same area but has hairy stems and leaves and flowers borne along an elongated stalk.

STEVE R. TURNER

AMERICAN GINSENG

Panax quinquefolius
Ginseng family (Araliaceae)

Description: Perennial, 6"–18" long, from an aromatic fleshy and forked root. Generally, 3 long-stalked leaves arise from the base and have compound leaves. Leaflets, 3–5, are up to 5" long, elliptical to inversely club-shaped, toothed along the margins, and pointed at the tip. Two of the leaflets are smaller than the other 3. Tiny greenish-yellow or greenish-white flowers are borne in ¾"-wide flat clusters arising from a leaf axil. Flowers are ⅛" wide and have 5 petals and 5 exserted stamens. Fruit is a reddish berry.

Bloom season: Summer

Range/habitat: Eastern North America and portions of the central United States in deciduous woodlands

Comments: *Panax* is from the Greek *panakeia* ("cure-all") after the plant's medicinal properties. *Quinquefolius* ("five-leaved") refers to the compound leaves. The leaves and roots were used by Native Americans for herbal treatments. The plants contain ginsenosides, compounds that are used medicinally to treat diabetic issues of blood sugar or insulin levels. In certain US locations, American ginseng has become rare due to overharvesting. Not to be confused with Oriental ginseng (*P. ginseng*), American ginseng is also commercially grown for its roots.

STEVE R. TURNER

WOOLLY DUTCHMAN'S PIPE

Isotrema tomentosa

Birthwort family (Aristolochiaceae)

Description: Perennial vine, 20'–30' long, but may reach 100'. Leaves, stems, and flowers are covered with fine, soft hairs. Heart- to kidney-shaped leaves are 4"–8" long and crowded along the stem. Greenish-yellow, 2"-long flowers, often hidden beneath the leaves, are curved and trumpet-shaped, resembling a Dutchman's smoking pipe. The odd-looking flower lacks petals; the calyx forms 3 lobes at the opening and is yellow with a purple center. Fruits are tubular capsules, about 3" long, which become gray-green with age.

Bloom season: Early spring to midsummer

Range/habitat: Southeastern and south-central United States along streams and in moist woods

Comments: *Isotrema* may refer to the 3 lobes of the style, which are equal in length. *Tomentosa* ("matted wool or short hair") refers to the soft white hairs on the leaves' undersides. Hairs inside the flowers point downward to prevent pollinating insects from escaping. Plants were once used to aid in childbirth. This is a larval host plant for the pipevine swallowtail (*Battus philenor*) butterfly.

STEVE R. TURNER

BLUE COHOSH

Caulophyllum thalictroides

Barberry family (Berberidaceae)

Description: Perennial, 1'–3' tall. Central stem is stout and greenish to light purple. Nonflowering plants have a single compound leaf that arises from this stem; flowering plant has 2 compound leaves that arise from the stem; leaves are bluish green. The lower compound leaf divides into 3 compound leaflets, which are again divided into 9 sub-leaflets arranged in groups of 3. Sub-leaflets are 1"–3" long, broadly egg-shaped, with 2–5 blunt lobes. The upper compound leaf is similar, but the 3 compound leaves are smaller and bear only 3 leaflets. Flowering clusters bear 5–30 yellow-green to purplish-brown flowers that are ⅓" wide and have 6 petallike sepals and insignificant petals. Six stamens surround a greenish center. Beneath each flower are 3–4 sepal-like bracts. Fruit is a green, berrylike seed with a fleshy coating that turns blue as it matures.

Bloom season: Mid-spring to early summer

Range/habitat: Eastern North America in moist sites in deciduous forests

Comments: *Caulophyllum* is from the Greek *kaulos* ("stem") and *phyllum* ("leaf"), in reference to the leaves. *Thalictroides* refers to the leaves resembling those of meadow rue (*Thalictrum* sp.). Seeds are toxic but have been reported as a coffee substitute. Flowers often appear before the leaves fully develop.

PHOTOGRAPHER DECLINED PHOTO CREDIT

CARDINAL AIRPLANT

Tillandsia fasciculata
Bromeliad family (Bromeliaceae)

Description: Epiphyte, often growing in dense clusters on trees. Greenish-gray leaves are 12"–30" long, leathery, and narrow, with a tapered and pointed tip. A 1'- to 2'-long flowering stalk arises from the leaves and bears several side branches bearing 3"- to 6"-long spikes of 10–50 flowers. Floral bracts are often red, although they may be green, yellow, or rose and overlap like scales. Three green sepals surround the 3, 1¾"-long petals, which are usually violet but occasionally white. Fruit is a capsule with tufted seeds.

Bloom season: Spring to early summer, but blooms year-round

Range/habitat: Southeastern United States and Mexico to northern South America and the West Indies, growing on trees and rocks in cypress swamps and pine woodlands

Comments: *Tillandsia* honors Elias Tillandz (originally Tillander) (1640–1692), a Swedish doctor and botanist who collected plants in Finland and utilized the medicinal value of plants in his patients' treatment. *Fasciculata* ("in a fascicle or little bundle") refers to the plant's basal cluster of leaves. The introduced Mexican bromeliad weevil (*Metamasius callizona*), habitat loss, and illegal collecting have severely impacted this airplant's distribution and survival. Airplants, or epiphytes, receive water and nutrients from the air, rainwater, and their host tree. Also known as giant airplant and dog-drink-water. Giant wild pine (*T. utriculata*) has leaves that are nearly 2' long and a flowering stalk nearly 6' long.

BORREGOWILDFLOWERS.ORG

TEDDY BEAR CHOLLA

Cylindropuntia bigelovii
Cactus family (Cactaceae)

Description: Perennial cactus, 1'–5' tall, covered with 1"-long silvery-white spines of equal length. The thick trunk produces lateral cylindrical stems, which may break off. Greenish-white flowers are about 1½" long and often have red patches on the petal tips. Fruit is ½"–1" long, fleshy, and many-seeded.

Bloom season: Early spring to early summer; possibly again in early fall

Range/habitat: Northwestern Mexico and the southwestern United States in dry desert chaparral

Comments: *Cylindropuntia* is from the Greek *kylindros* ("a cylinder"); *opuntia* is from a Greek word referring to a spiny plant growing in Opus, Greece. *Bigelovii* honors Dr. John M. Bigelow (1804–1878), a surgeon and botanist who served on a Mexican Boundary Survey and Pacific Railroad Survey of 1853–1854. Also known as jumping teddy bear for the broken-off stems that seem to "jump" and adhere to the skin or fur of a passerby. Microscopic barbs hold the stem in place before it is removed. These stem pieces may then root and start a new plant. Cactus wrens may be observed nesting in the upright stems, and woodrats haul broken stem pieces back to their burrows as protection against predators.

STEVE R. TURNER

WILD CUCUMBER

Echinocystis lobata
Gourd family (Cucurbitaceae)

Description: Annual vine; angular stems may reach up to 20' long. Coiled or straight tendrils arise on the stem and wrap around other plants for support. Alternate leaves are 6" wide, borne on long stems, and are palmately lobed, resembling a maple leaf with 5–7 pointed lobes. Separate male and female flowers are found on the plant. The white or greenish-yellow male flowers arise in branched clusters and have 6 slender lobes and a single stamen. Female flowers arise on short stems from the leaf axils and consist of a single style. Fruit is spiny and an oval-shaped, fleshy berry with a thick skin that resembles a cucumber.

Bloom season: Summer and fall

Range/habitat: Across much of North America along stream banks and in moist woods

Comments: *Echinocystis* is from the Greek *echino* ("hedgehog") and *cystis* ("bladder"), referring to the prickly fruit, known as a "pepo": a modified berry with a thick rind. As the fruits dry out, the seeds eventually drop to the ground from the bottom of the fruit. *Lobata* ("lobed) may refer to the interior lobed structure of the fruit. The plant has been used medicinally by Native Americans. Another common name is bur cucumber, for the prickly fruit.

STEVE R. TURNER

WILD YAM

Dioscorea villosa
Yam family (Dioscoreaceae)

Description: Highly variable twining vine, 5'–30' long, with rounded or angular stems. Heart-shaped leaves are mostly alternate (some basal leaves may be opposite or whorled), 2"–4" long, and palmately veined. Undersides of the leaves may have some hairs, and the 1½"- to 6"-long leaf stems may be light green or red and sometimes have hairs where the stem meets the leaf blade. Saucer-shaped male and female flowers are found on separate plants. Male flowers, which are about ⅛" wide with 6 whitish-green to yellowish-green tepals and 6 stamens, are arranged in clusters of 1–3 flowers along the branched flowering stalk. Female plants also have ⅛"-wide whitish-green to yellowish-green tepals that sit atop a large ovary with 3 wings. Female flowers are borne along an unbranched elongated flowering stalk. Fruit is a 1"-long ovoid capsule.

Bloom season: Mid-spring to midsummer

Range/habitat: Eastern North America in disturbed sites, thickets, woodlands, and sandy or rocky soils

Comments: *Dioscorea* is for Pedanius Dioscorides, a first-century Greek physician, botanist, and herbalist who wrote a five-volume Greek pharmacopeia on medicinal plants: *De Materia Medica. Villosa* means "with fine hairs." Although the plant does not produce edible tubers, the roots contain a compound called diosgenin and have been used medicinally for a variety of ailments; however, some sites warn of toxicity. The flattened seeds have membranous wings that help them get dispersed by the wind.

ELEANOR DIETRICH

VENUS FLYTRAP

Dionaea muscipula
Sundew family (Droseraceae)

Description: Carnivorous perennial; stem is 2"–5" tall. Leaves arise in a basal rosette from a short stem that resembles a bulblike base. The larger leaves with a long petiole that ends with a 2-lobed hinged appendage (the trap) form after the plant flowers. Appendages are fringed with long projections, and the inner surface has fine hairs. White flowers arise on a leafless flowering stalk about 6" long and have 5 petals with purplish nectar lines. Fruit is a cluster of black seeds.

Bloom season: Spring

Range/habitat: North and South Carolina in bogs, swamps, and wetlands in nitrogen-poor soils. Naturalized populations occur in Florida, New Jersey, and Washington, DC.

Comments: *Dionaea* is from the Greek meaning "daughter of Dione." Dione's daughter was Aphrodite, the goddess of love, who is similar to the Roman goddess Venus. *Muscipula* ("mousetrap" or "flytrap") describes the action of the leaves. These traps can determine the difference between live prey and debris or rain through the sensitive hairs located on the trap. Ants, spiders, beetles, and grasshoppers are the main prey species, with other flying insects representing a small portion of the diet. Digestive enzymes are released once the prey is securely trapped; complete digestion takes about 10 days. North Carolina designated the Venus flytrap as the state's carnivorous plant in 2005.

JIM FOWLER

ROUND-LEAF SUNDEW

Drosera rotundifolia
Sundew family (Droseraceae)

Description: Carnivorous perennial. Rounded clusters are up to 2"wide, with flowering stems 2"–10" tall. From a basal rosette, the leaves arise on ½"- to 2"-long hairy stems that terminate in a rounded leaf blade covered with long, reddish glandular hairs along the margins and shorter hairs in the blade's center. The long hairs have a large, "dewdrop-looking" gland, which insects become stuck to. The flowering stalk arises from the basal leaves and bears ⅓"- to ½"-wide white or pinkish flowers along one side of the stem. Fruit is a ½"-long tapered capsule.

Bloom season: Summer

Range/habitat: Circumboreal; found across much of North America in wetlands, bogs, stream edges, and fens

Comments: *Drosera* is from a Greek word meaning "dewy," in reference to the large, sticky drops that trap insects on the leaves. *Rotundifolia* ("round leaved") refers to the round leaf blade. Round-leaf sundew leaves have some of the highest concentrations of vitamin C of any plants. When an insect is stuck on the long hairs, these hairs curl inward as the leaf folds and the insect comes in contact with the smaller center hairs, which secrete enzymes that digest the prey.

STEVE R. TURNER

MONUMENT PLANT

Frasera speciosa

Gentian family (Gentianaceae)

Description: Monocarpic (blooms once then dies), often 4'–7' tall. Basal leaves are spatula-shaped or elliptical, 8"–20" long, and smooth or slightly hairy. Stem leaves are smaller and lance-shaped or inversely lance-shaped and arranged in whorls around the main stem. Flowers are borne in whorled clusters; each flower is ⅜"–¾" wide and has purplish dots on greenish petals and 2 glands on each petal's lobe. Fruit is a capsule.

Bloom season: Late spring through summer

Range/habitat: Western United States in shrub steppe, mountain meadows, and mountain brush

Comments: *Frasera* honors John Fraser (1756–1811), a Scottish botanist and explorer who traveled extensively and collected plants in America, Russia, Canada, and Cuba in the eighteenth century. He introduced more than 200 plants into European gardens. *Speciosa* ("showy") refers to the flowers. Recent research has determined that these plants bloom once during their 20- to 80-year life span and then die. The common name refers to the plant's stature. Also called showy gentian or elkweed.

CHECKER LILY

Fritillaria affinis

Lily family (Liliaceae)

Description: Perennial from a small fleshy-scaled bulb covered with rice-size bulblets. Unbranched stems reach 4"–40" tall. Lance-shaped linear leaves are borne in whorls at the base and in pairs along the upper stem. Bowl-shaped, mottled flowers, 1½" wide, have purple, yellow, or green mottling on the 6 tepals. Flowers hang downward, and the tepals flare open. Fruit is a many-seeded capsule.

Bloom season: Late spring to early summer

Range/habitat: Western North America in grasslands, meadows, and woodlands

Comments: *Fritillaria* ("dice box") refers to the seeds shaking around inside the capsule when mature. *Affinis* ("similar to") refers to the similarity of the flowers to other *Fritillaria* species. Flowers have a somewhat foul aroma. Also called mission bells or rice-root lily for the small bulblets attached to the roots.

CALIFORNIA CORN LILY

Veratrum californicum
Bunchflower family (Melanthiaceae)

Description: Perennial, growing 3'–8' tall from thick rhizomes. Often grows in dense clusters like a corn field. Large, overlapping oval to egg-shaped leaves are 8"–15" long and up to 8" wide. Small flowers are arranged in elongated clusters, up to 2' long, that branch upward from the main stalk. Clusters may contain hundreds of flowers. Individual flowers are 1" wide and have 6 whitish to greenish tepals, with a Y-shaped gland located at the base of each tepal. Mature capsule contains numerous winged, papery seeds.

Bloom season: Summer

Range/habitat: Western North America in moist and wet meadows, stream banks, and along rivers

Comments: *Veratrum* is from the Latin words *vere* ("true") and *ater* ("black"), for the black roots of another species. *Californicum* ("of California") is in reference to where the first specimen was collected for science. Toxic to livestock and humans. At one time an insecticide was made from the plant's powdered roots. Though the plants may produce few seeds some years, other years the plants will produce an abundance of seeds—heavy seeding known as mast seeding.

WILLIAM MCFARLAND

VIRGINIA BUNCHFLOWER

Veratrum virginicum
Bunchflower family (Melanthiaceae)

Description: Perennial, 2½'–5' tall with a stout stem. Basal linear leaves are 10"–20" long and have strong parallel venation. Stem leaves are similar but smaller; upper stem and flower branches bear curly hairs. Stem terminates in an open cluster of flowers that is 9"–18" long. Greenish-white flowers have 6 tepals, are ¾"–1" wide, and have 6 stamens. At each tepal's narrow base, there are 2 yellow, white, or green nectar glands. Fruit is a narrow, ¾"-long capsule.

Bloom season: Late spring to summer

Range/habitat: Eastern United States in swamps, damp prairies, springs, bogs, moist woods, fens, roadside ditches, and open bottomlands

Comments: *Veratrum* is from the Latin words *vere* ("true") and *ater* ("black"), for the black roots of another species. *Virginicum* ("of Virginia") refers to the distribution. The common name refers to the bunches of flowers that bloom. Contains toxic alkaloids. Also known as *Melanthium virginicum*.

CHUCK TAGUE

FLORIDA BUTTERFLY ORCHID

Encyclia tampensis
Orchid family (Orchidaceae)

Description: Epiphyte, from a 1"-wide pseudobulb. One to 3 alternate grasslike leaves, up to 6"–12" long, emerge from the pseudobulb. Open flowering cluster is borne on branching stems and may include up to 45 individual flowers that are 1"–1½" wide. Sepals and petals vary in color from green to yellow or bronze and surround a lobed lip that is ½"–¾" long. The labellum is lobed, not pouch-like, and white with purple spotting. The fragrant flowers smell like honey. Fruit is a capsule.

Bloom season: Late spring through summer

Range/habitat: Cuba, Bahamas, Florida, and southern Georgia, growing on trees in forests, swamps, and woodlands, including southern live oak, pond apple, red maple, gum, bald cypress, buttonwood, and pop ash

Description: *Encyclia* is from the Greek *enkykleoma* ("to encircle"), which refers to the lateral lobes surrounding the column. *Tampensis* ("Tampa") refers to the location of the type specimen. Pollination is often carried out by the sand wasp (*Rubrica nasuta*). The common name refers to the distribution of the plant and the fact that the flowers resemble butterflies when caught in a breeze.

STEVE R. TURNER

GIANT HELLEBORINE

Epipactis gigantea
Orchid family (Orchidaceae)

Description: Perennial from spreading underground roots. Upright stems are 1'–2', but may reach 5' tall. Stems bear numerous oval to lance-shaped leaves that are 2"–8" long. Greenish to purplish flowers have 3½"-long, greenish sepals and 3 petals with purple or dull red lines. The 2 upper petals are smaller, flat, and greenish purple; the lower petal is pouch-like and pinched in the middle to form a 3-lobed tip with a curled margin. Fruit is a 1"- to 1½"-long capsule that hangs downward.

Bloom season: Mid-spring to midsummer

Range/habitat: Western North America in seeps, springs, stream banks, and other wet meadows

Comments: *Epipactis* is from the ancient Greek name for the plant: *epipaktis* ("helleborine"). *Gigantea* ("large") refers to the stature of the plant. Wasps, attracted to the plant by scent, are the primary pollinators of these orchids; however, the flowers may self-pollinate. When touched, the flower's lower lip and tongue move, giving this plant another common name: chatterbox.

JIM FOWLER

CRESTED CORALROOT

Hexalectris spicata
Orchid family (Orchidaceae)

Description: Saprophytic or myco-heterotrophic plant, 6"–18" tall. Flowering stem rises up from the top of the taproot and is light brown to yellowish in color; there are no leaves. The stalk bears about 25 yellow-brown to purplish flowers. Flowers are made up of 3 sepals, light brown and streaked with purple, which spread outward and upward, and 3 petals. Two of the petals are similar; the lower petal, which forms the lip, is tan or magenta, 3-lobed, with wavy margins. Fruit is a capsule.

Bloom season: Mid-spring to late summer

Range/habitat: South-central and southeastern United States in a variety of habitats, including swamps, desert canyons, or woodlands in sandstone or limestone areas

Comments: *Hexalectris* is from Greek words meaning "six cocks," referring to the 6 raised ridges on the flower's lip that resemble a rooster's crest. *Spicata* ("a spike") refers to the arrangement of the flowers along an elongated stalk. The twisted roots resemble coral; hence the common name. These orchids cannot self-pollinate and attract bumblebees as their primary pollinator.

JIM FOWLER

GREEN ADDER'S-MOUTH ORCHID

Malaxis unifolia
Orchid family (Orchidaceae)

Description: Perennial, 4"–11" tall with a ribbed stem. A single (sometimes 2) oval to egg-shaped leaf has a pointed tip and wraps around the stem. Leaf is 1"–3" long and about half as wide. Off a main flowering stalk arise ⅜"-long twisted stems that terminate in a single flower. The 2-lipped green flowers are about ⅛" long. The upper 3 sepals flare outward, as do the 2 linear and recurved petals. The lower lip is 3-lobed. The flowering stalk may have up to 160 flowers. Fruit is a capsule.

Bloom season: Spring to fall

Range/habitat: Eastern and central North America, Mexico, Central America, and the Greater Antilles in bogs, swamps, and open woods

Comments: *Malaxis* ("smooth") refers to the texture of the leaves. *Unifolia* ("one-leaved") refers to the plant's single leaf. Mosquitoes, fungus gnats, and parasitic wasps are known pollinators of these plants.

JIM FOWLER

CRANE-FLY ORCHID

Tipularia discolor
Orchid family (Orchidaceae)

Description: Perennial, flowering stem, 15"–20" tall, arises from a corm. The single, egg-shaped leaf, 3"–4" long, arises in the fall and disappears in late spring or summer prior to flowering. Upper side of the leaf is shiny green to greenish purple, possibly with raised purple spots, and underside is purple. Leafless, greenish-brown flowering stalk bears 24–55 spindly, greenish-brown to coppery asymmetrical flowers, with the upper sepals and lateral petals pushed to one side. The lower lip is lobed and forms a nectar spur. Fruit is a capsule.

Bloom season: Late summer to early fall

Range/habitat: Eastern and central United States along stream edges and woodlands

Comments: *Tipularia* is a shortened version of *Tipula*, the genus of crane flies, which have long, delicate legs. *Discolor* ("of two colors") is in reference to the leaves. Noctuid or owlet moths are the primary pollinators of these flowers. As the moths probe the flower for nectar, they stick their proboscis deeper into the nectar spur. When the sticky pollen masses come in contact with the moth's compound eye, the transfer from flower to moth happens. The edible corms are potato-like. Also called crippled crane-fly, because the flowers resemble an injured crane fly.

MATT BERGER

CALIFORNIA GROUND-CONE

Kopsiopsis strobilacea
Broomrape/Paintbrush family (Orobanchaceae)

Description: Perennial, 4"–12" tall, cone-shaped. Parasitic on manzanita (*Arctostaphylos* sp.) or madrone (*Arbutus menzesii*). Overall color is reddish brown to purple. Lower oval-shaped, overlapping bracts are ½"–¾" long and pale edged, resembling the scales of a pinecone. Tiny purple flowers with pale edges peek out from between the scales and are 2-lipped; the upper lobe is hoodlike. Fruit is a small seed.

Bloom season: Late spring to midsummer

Range/habitat: Western United States in open forests or chaparral

Comments: *Kopsiopsis* ("like or having the form of *Kopsia*") honors Jan Kops (1765–1849), a Dutch professor of botany and an agronomist. *Strobilacea* ("cone-like fruits") refers to the aboveground structure. This plant spends about 9 months underground before just the flowering portion emerges above the ground. Specialized roots called haustoria connect the ground-cone to a host plant from which the *Kopsiopsis* draws nutrients and water. Since the ground-cone doesn't kill its host, it is a holoparasite. Another species, Vancouver ground-cone (*K. hookeri*) is similar. Some authors have kept these plants in the *Boschniakia* genus.

DESERT TRUMPET

Eriogonum inflatum
Buckwheat family (Polygonaceae)

Description: Annual or perennial, mostly 1'–3' tall. Smooth, greenish or brownish hollow stems are inflated near joints. Basal leaves are long-stalked, with rounded blades up to 1¼" wide with wavy margins. The many-branched flowering stalks arise from a common point; minute, yellowish or reddish petalless flowers are borne along slender branches. Fruit is a seed.

Bloom season: Mid-spring through summer

Range/habitat: Southwestern United States and Baja California in open areas, desert shrub, and pinyon-juniper woodlands

Comments: *Eriogonum* ("woolly knee") refers to the hairs located at the swollen joints of the stem, which may persist throughout the winter. *Inflatum* ("inflated") refers to the swollen stems. The inflated stems were long thought to be caused by a gall response to a female wasp (in the *Odynerus* genus) laying her eggs in the stems; however, recent research points to excessive buildup of carbon dioxide in the stem. Insects may use the inflated stem to deposit their eggs, but this is not the cause of the inflation. Native Americans ate the young stems raw or cooked.

MOUNTAIN-SORREL

Oxyria digyna
Buckwheat family (Polygonaceae)

Description: Perennial, 4"–20" tall, usually in dense tufts. Basal leaves are fleshy, round to heart-shaped, and borne on long stems. Leaves have wavy edges. The unbranched flowering stems rise above the leaves and bear numerous tiny green flowers (turning red with age) in dense clusters. Flowers lack petals but have 2 rows of segments that protect the 6 stamens and fused pistil with 2 styles. Fruit is a seed surrounded by a reddish wing.

Bloom season: Mid- to late summer

Range/habitat: Northern North America and western United States in rocky alpine soils, meadows, scree, or rock crevices in subalpine or alpine environments

Comments: *Oxyria* ("sour") refers to the taste of the tart, edible leaves due to the presence of oxalic acid. Mountain-sorrel is high in vitamin C. *Digyna* ("two women") refers to the flower's 2 fused carpels. As the plants emerge after the snow melts, the reddish leaves and stems will eventually turn green. The flowers are wind pollinated.

SAND DOCK

Rumex venosus

Buckwheat family (Polygonaceae)

Description: Perennial, stems 4"–20" tall. Stem leaves are lance- to egg-shaped, ¾"–5½" long, and ½"–2" wide. The numerous flowers have greenish floral segments that are ⅛" long. Flowers have 6 sepals, 6 stamens, and 1 pistil. The 3 inner sepals become enlarged when mature and form a seed with reddish wings.

Bloom season: Mid-spring to midsummer

Range/habitat: Central and western United States in dunes and sandy sites

Comments: *Rumex* is the Latin name for sorrel or dock. *Venosus* ("veined") refers to the prominent leaf veins. The common name refers to the type of habitat the plant grows in. Boiled roots produce red, yellow, or black dyes. A poultice from the mashed roots has been used to treat burns. Though the leaves and shoots are considered edible, they contain toxic oxalates and should be boiled to render the oxalates harmless. Canaigre, or wild rhubarb (*R. hymenosepalus*), has large basal leaves and winged seeds. The roots were used in tanneries to dye leather.

CALIFORNIA PITCHER-PLANT

Darlingtonia californica

Pitcher Plant family (Sarraceniaceae)

Description: Perennial, insectivorous plant. Large leaves, 5"–20" long, are hood-shaped and greenish yellow with purple spots on the hood. At the base of the hood is an opening with 2 moustache-like bracts. A single, 2"-long flower is borne on a leafless stalk that is 1'–3' tall. The 1"- to 2½"-long flowers have cream sepals and purplish petals; flowers hang downward. Fruit is a turban-shaped capsule.

Bloom season: Mid-spring through summer

Range/habitat: Western United States in Oregon and northern California in nitrogen-poor soils such as bogs, springs, fens, wetlands, and along small streams

Comments: *Darlingtonia* honors Dr. William Darlington (1782–1863), a Philadelphia botanist. *Californica* ("of California") refers to its type location. Flies, wasps, bees, ants, and beetles are attracted to the nectar source within the flower opening. However, as they try to leave, the opaque spots on the hood "confuse" the insect, which spends a lot of time and energy trying to escape. As the insect tires, it falls into the digestive fluid at the leaf base and is digested by enzymes. Also known as cobra lily, for the leaves resembling a rising cobra.

FRAGRANT FRINGECUP

Tellima grandiflora
Saxifrage family (Saxifragaceae)

Description: Perennial, 1'–3½' tall. Basal leaves are heart-shaped and 2"–4" wide, with 5–7 toothed, shallow lobes. The few stem leaves are similar but smaller. Tubular, bell-shaped flowers have 10 stamens and are borne in loose clusters along a flowering stalk. Petals are frilled and greenish white or pink, turning dark red with age. Fruit is a capsule.

Bloom season: Late spring to midsummer

Range/habitat: Western North America in moist woods and along stream banks, avalanche tracks, and mountain slopes

Comments: *Tellima* is an anagram of *Mitella*, another genus in the Saxifrage family. *Grandiflora* ("large-flowered") refers to the sizable flowers; the common name refers to the fringe around the rim of the flower. Native Northwest Indigenous peoples made an infusion of the crushed leaves to treat stomach issues. Also called false alumroot because the leaves resemble those of alumroot.

STEVE R. TURNER

BLUE RIDGE CARRION FLOWER

Smilax lasioneura
Greenbriar family (Smilacaceae)

Description: Perennial vine that dies back each year; grows to 7' tall. Stems are thornless and have tendrils. The egg-shaped leaf blade arises on a short stem, is 1½"–3" long, hairy below and dull green above. Small green flowers are borne in a dense, rounded cluster. Fruit is a ½"-wide blue-black berry.

Bloom season: Spring to summer

Range/habitat: Central and eastern North America in deciduous woods

Comments: *Smilax* ("clasping") refers to a Greek myth about the tragic affair between Crocus, a mortal man, and the nymph Smilax, who is turned into a vine while the human is turned into the crocus flower. *Lasioneura* ("woolly veins") refers to the fine hairs on the underside of the leaves arising along the leaf veins. The foul odor of the flowers gives the plant its common name and attracts flies, beetles, and bees as pollinators. The young shoots and leaves are edible, as are the berries. All other *Smilax* species in the eastern United States have thorny stems.

STEVE R. TURNER

GIANT BUR-HEAD

Sparganium eurycarpum

Bur-reed family (Sparganiaceae)

Description: Perennial, 2'–6' tall. Long stiff linear leaves are 1'–5' long, have entire margins, and are keeled along the lower portion and flat toward the tip. The branched flowering stalk bears up to 6 female flower heads and up to 20 male flower heads. Male flowers are located above the female flowers, and the flowering stalk often bends where a female flower cluster arises. Female flowers are densely clustered in a ¾"- to 1½"-wide, globe-shaped, prickly, greenish flower head. There are 2 styles for each flower. Male flowers are also densely packed into a cluster that is smaller than the female's flower head and whitish when the flowers open. Each flower has 5 white stamens. Fruit is a pyramid-shaped seed with a beaked tip at the base.

Bloom season: Summer

Range/habitat: Most of North America except the southeastern United States in wetlands, fens, marshes, ponds, and along slow-moving rivers

Comments: *Sparganium* is from the Greek word *sparganon* ("swaddling band") in reference to the strap-like leaves. *Eurycarpum* ("broad-fruited") refers to the seeds. This bur-head often grows in areas that experience seasonal flooding, and the plants can regenerate through seeds or from sprouting along the rhizomes. The seeds are an important food source for waterfowl, and muskrats often eat the plant's rhizomes. The flowers are wind pollinated.

STINGING NETTLE

Urtica dioica

Nettle family (Urticaceae)

Description: Perennial from creeping roots, with stems 3'–9' tall. Stems and leaves are covered with stinging hairs. Leaves are opposite, broadly lance- to egg-shaped, with saw-toothed margins. Clusters of tiny, nondescript greenish flowers are borne in the upper leaf axils. Flowers may be all male, all female, or have both male and female appendages. Fruit is a flat, oval seed.

Bloom season: Summer

Range/habitat: Western North America in moist sites along streams, rivers, thickets, and woodlands

Comments: *Uro* ("to burn") is the Latin name for nettle. *Dioica* ("dioecious") refers to the male and female flowers, which are borne on separate plants. The stem fibers were plaited into cordage by Indigenous peoples. The cooked stems and leaves were, and still are, eaten as "wild spinach." The hollow hairs along the stems and leaves contain formic acid, which irritates the skin.

ORANGE FLOWERS

This section includes orange flowers as well as multicolored flowers that are predominately orange. Since orange flowers often become either paler or deeper in color with age, you should check both the "Yellow Flowers" and "Red and Violet Flowers" sections if you do not find the flower you are looking for in this section.

KATIE BYERLY

BUTTERFLY MILKWEED

Asclepias tuberosa
Dogbane family (Apocynaceae)

Description: Perennial, up to 3' tall, with stout stems covered in small, coarse hairs. When broken, the leaves and stem bleed a clear sap. Narrow, short-stemmed leaves are lance- to dagger-shaped and approximately 4" long. Lower leaves are alternate; upper leaves may be opposite. Flower clusters have about 25 short-stalked, orange to yellowish-red flowers, each with 5 reflexed petals and 5 erect horns. Spindle-shaped pods, up to 6" long, have small, soft hairs.

Bloom season: Late spring to midsummer

Range/habitat: Mostly central and eastern North America; also parts of the southwestern United States and California in sagebrush, grasslands, prairies, and woodlands

Comments: *Asclepias* refers to Asklepios, a mortal Greek physician who was an authority on plants and their healing properties. According to Greek mythology, Asklepios was killed by a thunderbolt from Zeus after Asklepios boasted about being able to revive the dead. *Tuberosa* ("swollen") refers to the thickened roots, which are used to treat certain lung ailments; another common name for this plant is pleurisy root. Indigenous peoples plaited the stems' strong fibers into cordage. Unlike many other milkweed species, butterfly milkweed does not have a milky sap. When in bloom, various insects, including butterflies—hence its common name—are attracted as pollinators.

MATT BERGER

ORANGE GLANDWEED

Adenophyllum cooperi
Aster family (Asteraceae)

Description: Perennial, 1'–2' tall; stems and leaves have fine hairs. Leaves are 2½"–8", alternate, oval-shaped or wider at the top and narrower at the base, and lobed near the base and irregularly toothed along the edge. Leaves also have 1 or 2 pairs of yellow oil glands at the base and tip. The orange to yellow-red flower heads are subtended by up to 22 linear bracts that taper to a tip; numerous small disk flowers may be surrounded by up to 13 ray flowers, but ray flowers may be absent.

Bloom season: Late spring to late summer

Range/habitat: Southwestern United States in desert washes, sandy sites, and alluvial plains

Comments: *Adenophyllum* ("glandular leaf") refers to the oil glands located on the leaves. *Cooperi* honors either William Cooper (1798–1864), an American naturalist, or his son, James Graham Cooper (1830–1902), an American surgeon and naturalist who collected birds and plants in the western United States. The flowers may bloom a second time in the fall if conditions are favorable. Also called Cooper's dogweed.

ORANGE AGOSERIS

Agoseris aurantiaca
Aster family (Asteraceae)

Description: Perennial, 4"–24" tall. Basal leaves are 2"–12" long, lance- to egg-shaped, pointed at the tip, and generally hairless but may have some soft hairs. The flowering stalk bears a single, ½"- to ¾"-wide flowering head with orange to orange-red ray flowers with square-tipped petals. Bracts beneath the flowering head are upright, not reflexed. Fruit is a beaked seed topped with white hairs.

Bloom season: Summer

Range/habitat: Western North America in mountain meadows, along stream banks, open woodlands, and mixed in mid- to high-elevation shrubs

Comments: *Agoseris* ("goat chicory") has a mysterious origin. Although bearing similar flowers to forage chicory (*Cichorium intybus*), which was feed to goats, this genus was named by eccentric European naturalist Constantine Samuel Rafinesque-Schmaltz (1783–1840), who left no clue as to why. *Aurantiaca* ("orange") describes the color of the flower heads, although the flowers may vary from rusty orange to pink. This plant resembles the common dandelion; hence one of its common names is orange mountain dandelion.

ORANGE SNEEZEWEED

Hymenoxys hoopesii
Aster family (Asteraceae)

Description: Perennial, 1'–3' tall. Oblong basal leaves are entire and hairy or smooth. Upper leaves are lance-shaped, lack a stem, are 3"–6" long, and have a nearly white central vein. Orange-yellow flower heads are borne on woolly stalks and have 14–26 drooping ray flowers, 1½" long, surrounding a center mound of orangish disk flowers. Fruit is a seed with scales.

Bloom season: Summer

Range/habitat: Western United States in moist meadows and open woods

Comments: *Hymenoxys* is from the Greek *hymen* ("membrane") and *oxys* ("sharp"), in reference to the pointed pappus scales. *Hoopesii* honors Thomas Hoopes (1834–1925), a farmer, civic leader, and amateur botanist who traveled west and collected plants and seeds in Colorado. Roots were used to treat stomach disorders and rheumatic pain. Also called owl's-claws, for the drooping petals.

KATIE BYERLY

ORANGE JEWELWEED

Impatiens capensis
Jewelweed family (Balsaminaceae)

Description: Annual, 1'–5' tall with succulent stems; grows in dense clumps. Alternate leaves are elliptical to oval and toothed along the margins. Leaves appear silvery under water. Flowers are 1"–2" long and orange- or brown-spotted. The horn-shaped flowers appear inflated, but there are 4 petals and 3 sepals, one of which forms a curved spur. Fruit is a capsule containing several seeds.

Bloom season: Summer

Range/habitat: Across much of North America except the southwestern United States, Montana, and Wyoming in moist sites along streams, rivers, and thickets

Comments: *Impatiens* ("impatient") refers to the dispersal mechanism of the seeds. When the ripe capsule is touched, the pods explode lengthwise—they are "impatient" with being touched—and hurl the seeds (known as "explosive dehiscence"). *Capensis* ("of the Cape of Good Hope, South Africa") refers to the resemblance of the flowers to the outline of South Africa. Native Americans crushed the succulent stems and applied the juice to poison ivy rashes. The common name may refer to the silver color of the leaves under water. Also called touch-me-not for the exploding capsule.

ALEX HEYMAN

DESERT MARIPOSA LILY

Calochortus kennedyi
Lily family (Liliaceae)

Description: Perennial, up to 23" tall but often shorter; stems may be twisted. Basal leaf (may be several) is grasslike, 4"–8" long, wavy, and withers prior to flowering. Bell-shaped orange to yellow-orange flowers are arranged in loose clusters of 1–6 flowers; flowers have 3 sepals and 3 petals. At the base of the 2"-long petals is a large, dark purplish patch partially obscured by reddish-purple yellow-tipped hairs. Fruit is an angled, striped 3"-long capsule.

Bloom season: Spring to early summer

Range/habitat: Southwestern United States and northern Mexico in grasslands, dry slopes, desert shrub, and Joshua tree or pinyon-juniper woodlands

Comments: *Mariposa* is Spanish for "butterfly." Calochortus ("beautiful grass") refers to the leaves. *Kennedyi* honors William Kennedy (c. 1827–?), an American plant collector. The petals may range from yellow to orange to vermillion in color. These plants do not flower each year.

COLUMBIA LILY

Lilium columbianum
Lily family (Liliaceae)

Description: Perennial, 1'–4' tall; stems usually hairless. Lance-shaped leaves are 2"–5" long, arranged in several whorls of 2–9 leaves. Upper stem leaves are variable. Showy orange flowers have red or purple spots near the center and hang downward. Tepals curve backward to expose the orange anthers. Fruit is a barrel-shaped capsule that contains numerous flat, stacked seeds.

Bloom season: Summer

Range/habitat: Western North America in meadows, along stream banks, clearings, or open forests

Comments: *Lilium* is from the Greek *leirion* ("a lily"). *Columbianum* ("Columbia River") refers to the area where this plant was first collected for science. The edible bulbs are peppery. Hummingbirds and bees pollinate the flowers, which may also be reddish or yellow-orange.

ELEANOR DIETRICH

TURK'S CAP LILY

Lilium superbum
Lily family (Liliaceae)

Description: Perennial, 3'–7' tall, with a stout stem from a white bulb. Elliptical to lance-shaped leaves are stiff, up to 7" long, and arranged in a whorled pattern of 3–9 around the stem. The nodding flowers are 2½"–4" wide, reddish-orange to orange, with reddish tips and a green center with maroon spots, and have recurved tepals. Stamens project beyond the flowers, and the dark anthers are about ½" long. Fruit is a capsule with stacked seeds with thin, papery wings.

Bloom season: Early to midsummer

Range/habitat: Eastern and central North America in wet meadows and moist woodlands

Comments: *Lilium* is from the Greek *leirion* ("a lily"). *Superbum* ("superb") refers to the flower. Indigenous peoples harvested the roots for food. The common name refers to the resemblance of the flowers to a type of hat worn by Turkish people. The papery wings on the seeds enable them to be dispersed by the wind. Michigan lily (*L. michiganense*) is similar but has yellow bulbs and lacks the green centers on the tepals.

SCARLET GLOBE MALLOW

Sphaeralcea coccinea
Mallow family (Malvaceae)

Description: Perennial, with stems solitary or many from a woody base. Plants are 2"–18" tall. Leaf blades are longer than wide; 3–5 deep lobes may be again lobed or toothed. Orange to scarlet flowers, ½" wide with numerous stamens, are borne along an elongated stalk. Fruit is a rounded capsule containing numerous tiny black seeds.

Bloom season: Mid-spring to midsummer

Range/habitat: Western North America in grasslands, shrublands, sandy sites, and ponderosa pine woodlands

Comments: *Sphaeralcea* is from the Latin *sphaira* ("globe") and *alcea* (a related genus) and refers to the rounded capsules resembling those of the *Alcea* genus. *Coccinea* ("scarlet") refers to the floral color. Sometimes bees in the genus *Diadaysia* can be found in the morning curled up inside a flower after having spent the night in their sleeping quarters.

JIM FOWLER

ORANGE FRINGED ORCHID

Platanthera ciliaris
Orchid family (Orchidaceae)

Description: Perennial, up to 2½' tall. Two to 4 lance-shaped leaves are alternately arranged and clasp the stem; upper leaves are reduced to bracts. The flowering stalk bears numerous, up to 115, orange to yellow flowers that have a lower lip with heavy fringe and a long, cylindrical spur that projects behind the flower, curving downward. Fruit is a capsule.

Bloom season: Early summer to early fall

Range/habitat: Central and eastern North America in prairies, meadows, bogs, marshes, and flatwoods in acidic soils

Comments: *Platanthera* ("broad anther") describes the shape of the anther. *Ciliaris* ("edged with hairs") describes the lower lip. Butterflies such as swallowtails are common pollinators of these flowers. Also known as yellow fringed orchid or fringed orange bog orchid, both of which indicate the fringed flowers and bog-loving habitat. Crested orange bog orchid (*P. cristata*) is similar but smaller.

MADELEINE CLAIRE

CALIFORNIA POPPY

Eschscholzia californica

Poppy family (Papaveraceae)

Description: Perennial, 6"–22" tall. Highly divided basal leaves have a bluish tint and a lacy appearance. Flowering stalks bear a single, orange to yellow-orange, bowl-shaped flower with 4 delicate petals that are ½"–2" long. Flower sits on a distinct rim that becomes more noticeable in fruit. Fruit is a long, slender seedpod.

Bloom season: Spring to fall

Range/habitat: Western United States to northern Mexico in grassy slopes, meadows, rock outcrops, and disturbed sites

Comments: *Eschscholzia* is named for Johann Friedrich von Eschscholtz (1793–1831), an eminent German physician, botanist, and naturalist who completed two circumnavigations of the globe with explorer Otto von Kotzebue, 1815–1818 and 1823–1826. *Californica* ("of California") refers to the location of the type specimen. The flowers close at night or during cloudy days. Bees, wasps, and beetles might be observed pollinating the flowers. A widely used ornamental that is also used in roadside restoration projects, this plant is California's state flower.

FIRE POPPY

Papaver californicum

Poppy family (Papaveraceae)

Description: Annual, up to 24" tall. Basal and stem leaves have fine hairs; lower leaves are up to 3½" long, pinnately lobed, with the side lobes again divided. The flowering stalk bears a single, 4-petaled orange to reddish-orange flower that is ½"–1½" wide with a green center. Bowl-shaped flower has about 20 stamens. Stems are bent over when the flower bud forms but become erect when the flower opens. Fruit is a capsule.

Bloom season: Mid-spring to midsummer

Range/habitat: Endemic to California in chapparal and oak woodlands

Comments: *Papaver* is the Latin term for "poppy." *Californicum* ("of California") describes the distribution of this plant. For many seeds, moisture is a trigger to germinate, but fire poppies are in a group of plants called "fire followers," which are triggered by fire, smoke, or charred soils to germinate. Smoke, which contains karrikin compounds, triggers germination in this species. The seeds may lie dormant for years.

LARGE-FLOWERED COLLOMIA

Collomia grandiflora

Phlox family (Polemoniaceae)

Description: Annual, stems up to 40". Plants may have 1 or many stems. Narrow leaves are up to 2" long and linear. The 1"-long funnel-shaped flowers are arranged in sticky clusters and are salmon, light red, yellow, or white in color. The 5 petals flare open to reveal (barely) the 5 stamens with blue anthers. Fruit is a capsule.

Bloom season: Late spring and summer

Range/habitat: Western North America in dry, open woodlands

Comments: *Collomia* is from the Greek *colla* ("glue") and refers to the stickiness of the wet seeds. *Grandiflora* ("large flower") refers to the flower's size. The roots and leaves were brewed by some Northwest tribes as an eyewash or laxative or to treat fevers. The anthers produce blue pollen, which is unique.

RED AND VIOLET FLOWERS

STEVE R. TURNER

This section includes red and violet flowers as well as multicolored flowers that are predominately red or violet. Since red and violet flowers often become either paler or deeper in color with age, you should check both the "Blue Flowers" and "Orange Flowers" sections if you do not find the flower you are looking for in this section.

ALEX HEYMAN

WESTERN SEA-PURSLANE

Sesuvium verrucosum

Ice Plant family (Aizoaceae)

Description: Perennial; sprawling prostrate stems up to 3' long. Leaves are variable in shape from linear to spatula-shaped, succulent, generally less than 2" long, and covered with crystalline bumps. Petalless flowers are rose-pink to orange to red in color, have 5 pointed sepals, and arise from the leaf axils. Around 30 pink stamens arise inside in the flower; the pistil is 5-lobed. Fruit is a capsule.

Bloom season: Spring to fall

Range/habitat: Western United States to South America in both coastal and inland salt marshes, wetlands, dry playas, and alkali flats

Comments: The meaning of *Sesuvium* ("the land of Sesuvii, a Gallic tribe") to these plants is unknown. *Verrucosum* ("wartlike") refers to the crystalline bumps located along the leaves and stems. On the eastern US coast, shoreline sea-purslane (*S. portulacastrum*) grows in sandy dunes and beaches. Both species have edible leaves and contain ecdysterone, a compound that controls molting for insects and crustaceans. Also known as ice plant.

STEVE R. TURNER

PRAIRIE ONION

Allium stellatum

Amaryllis family (Amaryllidaceae)

Description: Perennial; from a bulb and with stout, upright stems 12"–18" tall. Basal leaves are chive-like except solid, up to 12" long, and often wither prior to flowering. The reddish-pink to white flowers are borne in dense, rounded clusters that are 3"–4" wide. The starlike flowers have 3 petals and 3 sepals; flowers do not produce bulblets. Fruit is a capsule.

Bloom season: Midsummer to early fall

Range/habitat: Central North America in prairies, grasslands, rocky hillsides, and limestone-rich soils

Comments: *Allium* is the classical name for garlic. *Stellatum* ("starlike") refers to the flowers. The leaves and flowers have an oniony aroma, and the bulbs have been harvested for food, although their culinary value is low. Small bees and flies are common pollinators of these flowers. Also known as cliff or prairie onion due to its habit of growing on bluffs, rocky hillsides, and prairies.

STEVE R. TURNER

SWAMP MILKWEED

Asclepias incarnata
Dogbane family (Apocynaceae)

Description: Perennial, 2'–4' tall; stem and leaves bleed light milky latex when broken. Stem leaves are opposite, lance-shaped, and 3"–6" long. Flowering head at the end of the stalk is composed of numerous small, rose-colored flowers. The ¼"-wide flowers have 5 reflexed petals and an elevated central column. Fruit is a 4"-long, tan seedpod.

Bloom season: Summer

Range/habitat: Northeastern and southeastern United States in meadows, swamps, and river bottomlands

Comments: *Asclepias* refers to Asklepios, a legendary Greek physician who was an authority on the medicinal properties of plants. According to Greek mythology, Asklepios boasted that he could bring the dead back to life. Hades, god of the dead, took exception to this boast and coerced Zeus, Hades's brother, to kill Asklepios with a thunderbolt. *Incarnata* ("flesh colored") refers to the flower color. As is the case with many milkweeds, this is a host plant for monarch butterflies.

SHOWY MILKWEED

Asclepias speciosa
Dogbane family (Apocynaceae)

Description: Perennial, with stout stems that may be up to 3' or taller. Opposite leaves are broad, hairy, and bleed a milky latex when broken. Spherical clusters of pinkish-white flowers often hang downward from short, stout stalks. The 1"- to 1½"-long flowers are star-shaped and have 5 rose-purple reflexed petals and 5 pinkish-cream, needlelike, pouch-shaped hoods. Fruit is a 3"- to 5"-long, spiny or smooth pod that contains numerous seeds with long plumes.

Bloom season: Mid-spring through summer

Range/habitat: Western and central North America in disturbed areas along roads, ditches, waterways, and moist sagebrush shrublands

Comments: *Asclepias* refers to Asklepios, a legendary Greek physician who was an authority on the medicinal properties of plants. According to Greek mythology, Asklepios boasted that he could bring the dead back to life. Hades, god of the dead, took exception to this boast and coerced Zeus, Hades's brother, to kill Asklepios with a thunderbolt. *Speciosa* ("showy") is in reference to the spectacular flowers. The silky down attached to the seeds, which is 5 or 6 times more buoyant than cork, was used to stuff pillows and World War II–era life jackets and flight suits. Showy milkweed is a host plant for monarch butterfly larvae and a nectar source for many insects. The plant is toxic to livestock.

KATIE BYERLY

EASTERN SKUNK CABBAGE

Symplocarpus foetidus

Arum family (Araceae)

Description: Perennial, 1'–3' tall. Flowers appear before the leaves in early spring. A reddish-brown mottled, twisted shell (the spathe) envelopes a greenish-yellow flowering stalk (the spadix), which is 2"–4" long and egg-shaped. The sharply pointed spathe never opens fully. Individual flowers are ¼" long, lack petals, and have 4 fleshy sepals. As the flowers fade, large cabbage-like leaves grow that are 16"–22" long and 12"–16" wide, with pronounced venation. Fruit is a large seed.

Bloom season: Early spring

Range/habitat: Much of eastern North America in marshes, bogs, and wetland edges.

Comments: *Symplocarpus* is from the Greek *symplokos* ("connected") and *karpos* ("fruit"), referring to the clustered fruits. *Foetidus* ("foul-smelling") refers to the aroma of the leaves. Through a chemical metabolic process called cyanide resistant respiration, the plant can generate heat from 20°F to 60°F above air temperature and emerge through frozen ground or snow. The heat may also enhance the fetid aroma of the flowers to attract pollinators. When crushed or torn, the leaves release a putrid aroma that attracts early-season carrion-eating flies as pollinators. The plants contain oxalate crystals, which are considered toxic but may be rendered harmless through proper preparation. The pea-size seeds may be transported by water or in animal droppings. Also known as pole-cat weed or hermit of the bog.

STEVE R. TURNER

CANADIAN WILD GINGER

Asarum canadense

Birthwort family (Aristolochiaceae)

Description: Perennial, creeping along the ground, 4"–8" tall. Two heart- to kidney-shaped leaves, up to 6" wide and covered with fine, soft hairs, arise from the base on long slender stems. Cup-shaped flowers are reddish brown, 1" wide, and arise on short stems that trail along the ground. Flowers are borne singly on the stem and are composed of 3 fused sepals that have tapered tips. Leaves often obscure the flowers. Fruit is a pod.

Bloom season: Spring

Range/habitat: Central and eastern North America in wooded areas

Comments: *Asarum* is from the Greek and Latin names for the plant. *Canadense* ("of Canada") refers to the northern distribution of the plant. Unrelated to commercially harvested ginger (*Zingiber officinale*), the roots of wild ginger have a ginger-like aroma and have been used as a substitute for the commercial plant. In western North America, wild ginger (*A. caudatum*) is similar, but the flowers have longer, tapered points on the sepals. Both species of wild ginger have fleshy caps on the seeds called elaiosomes that are high in protein and attract ants, which help in dispersal by hauling the seeds to their mounds.

STEVE R. TURNER

PURPLE CONEFLOWER

Echinacea purpurea
Aster family (Asteraceae)

Description: Perennial, 1½'–4' tall, with branching stems. Egg- to lance-shaped basal leaves are 2"–6" long, hairy, and toothed along the margins. Large solitary flower heads have both ray and disk flowers and may reach 5" in diameter. Ten to 20 purple to pinkish ray flowers droop or spread somewhat horizontally and surround a dome-like central cluster of disk flowers. Fruit is a 3- or 4-angled seed.

Bloom season: Early to late summer

Range/habitat: Central United States in meadows, prairies, and open woodlands

Comments: *Echinacea* is from the Greek *echinos* ("hedgehog" or "sea urchin"), in reference to the spiny nature of the seed head. *Purpurea* ("purple") refers to the flower color. Flowers attract butterflies and finches as pollinators and seed dispersers, respectively. Native Americans used the plant to treat a variety of ailments, including toothaches, sore throats, and snakebites, and distemper in horses. Current pharmaceutical uses include boosting the immune system.

STEVE R. TURNER

SWEET JOE-PYE WEED

Eutrochium purpureum
Aster family (Asteraceae)

Description: Perennial, clump-forming, 3'–7' tall. Lance- to egg-shaped leaves arise from purple nodes along a light green stem in whorls of 3–4 leaves, are 6"–12" long and serrated along the margins. Leaves may have fine hairs on the undersides. Flower clusters are in rounded heads with numerous small pinkish-purple, vanilla-scented flowers. Disk flowers are arranged in small clusters arising on slightly hairy stems; ray flowers are absent. Disk flowers have 5 small teeth that rise above the rim; the bracts subtending the flowers are pinkish. Fruit is a bullet-shaped seed with fine hairs.

Bloom season: Summer

Range/habitat: Eastern and central North America in wet meadows, thickets, stream edges, and woodlands

Comments: *Eutrochium* is from the Greek *eu* ("well") and *troche* ("wheel-like"), in reference to the whorled arrangement of the leaves. *Purpureum* ("purple") refers to the flower color. This species prefers drier soils than other *Eutrochium* species. The common name is for Joseph Shauquethqueat, aka Joe Pye, an eighteenth- to nineteenth–century Mohican sachem.

TOM LEBSACK

DOTTED BLAZING STAR

Liatris punctata
Aster family (Asteraceae)

Description: Perennial, 6"–36" tall, with stems arising singularly or in clusters. Narrow leaves are basal, up to 5" long, with tiny hairs and tiny round dots along the margins. Pink to purple disk flowers are borne in small heads densely clustered along an elongated stalk. Flower heads have 4–8 disk flowers. Fruit is a seed with plumelike bristles.

Bloom season: Early summer to early fall

Range/habitat: Central Canada through the central United States into northern Mexico in a variety of habitats, including open fields, prairies, meadows, sagebrush shrublands, and ponderosa pine woodlands

Comments: The derivation of *Liatris* is uncertain. *Punctata* ("dotted") refers to the spotting on the leaves. The common name blazing star refers to the starlike pattern of the flowers. The flowers attract a variety of butterflies as pollinators, including a rare subspecies of Leonard's skipper (*Hesperia leonardus*) called the Pawnee montane skipper (*H. leonardus montana*). Native Americans used this plant as a food source. Scaly gayfeather (*L. squarrosa*) is similar and grows in the southeastern United States.

TOM LEBSACK

TALL BLAZING STAR

Liatris pycnostachya
Aster family (Asteraceae)

Description: Perennial, 2'–5' tall. Stems arise from a basal rosette of narrow, linear or lance-shaped leaves that may be up to 12" long. The small flower heads, about ¼"–¾" wide, are arranged in dense clusters along an elongated stalk that also bears numerous linear leaves smaller than the basal ones. The rose-purple disk flowers are arranged in groups of 5–10 per flower head; the heads are subtended by long, reddish bracts with curved tips. Fruit is a seed with light brown hairs.

Bloom season: Mid- to late summer

Range/habitat: Central and eastern United States in prairies, meadows, disturbed areas, and open woodlands

Comments: The derivation of *Liatris* is uncertain. *Pycnostachya* is from the Greek ("crowded"), which may refer to the arrangement of the leaves or flowers. The tall plants resemble fairy wands blowing in the breeze. Long-tongued bees and butterflies are common pollinators of these flowers. Also known as cat-tail or thick-spike gayfeather.

JUDY PERKINS

MOHAVE HOLE-IN-THE-SAND PLANT

Nicolletia occidentalis
Aster family (Asteraceae)

Description: Perennial, 5"–12" tall, with multiple thin stems. Fleshy linear to cylindrical leaves are ¾"–4" long, may have several lobes along the margins, and are bristle tipped. Flowering heads have 8–12 pale to bright pink ray flowers surrounding a center of yellow disk flowers. Fruit is a seed with bristles.

Bloom season: Mid-spring to early summer

Range/habitat: California and south into Baja California in sandy deserts

Comments: *Nicolletia* honors Jean-Nicolas Nicollet (1786–1843), a French cartographer and astronomer who mapped the Upper Mississippi River Basin in the 1830s. The Nicollet Tower stands in Sisseton, South Dakota, as tribute to his work. *Occidentalis* ("western") refers to the plant's distribution. The plants often arise from a depression in the desert, giving this plant its common name.

BORREGOWILDFLOWERS.ORG

SPANISH NEEDLE

Palafoxia arida
Aster family (Asteraceae)

Description: Annual, up to 6' tall; stems variable with glandular hairs or smooth. Linear to lance-shaped leaves are up to 5" long and have a rough, hairy texture. Cylindrical flowering heads contain up to 40 individual purple and white disk flowers that are ¾"–1" long and have long, pointed bracts with glandular hairs that subtend the heads. The 5-lobed flowers have a purple style and stamens that protrude beyond the mouth of the flower. Fruit is a seed with narrow pointed scales.

Bloom season: Late winter through spring

Range/habitat: Southwestern United States and northwestern Mexico in sandy sites such as dunes or desert washes

Comments: *Palafoxia* may honor José de Palafox y Melzi, Duke of Saragossa (1776–1847), a Spanish general who fought against Napoleon's army. However, recent research indicates that the genus may honor Juan de Palafox y Mendoza (1600–1659), a bishop and founder of the University of Mexico. *Arida* ("arid") refers to the habitat where this species grows. The common name comes from the pointed scales at the tip of the seedlike fruit, which resemble needles.

STEVE R. TURNER

PRAIRIE CONEFLOWER
Ratibida columnifera
Aster family (Asteraceae)

Description: Perennial, 1½'–3' tall. Lower leaves are pinnately lobed, up to 6" long; the 5–13 leaflets are linear. Flowering heads are composed of 4–11 yellow or yellow-brown drooping ray flowers surrounding a center of reddish-brown disk flowers that projects above the ray flowers by ½"–2". Ray flower petals may range in color from red to yellow or a mix of both colors.

Bloom season: Late spring to fall

Range/habitat: Across much of North America in prairies, meadows, savannas, and open plains

Comments: The meaning of *Ratibida* is a bit obscure. *Columnifera* ("column-like") refers to the column-like arrangement of the disk flowers. Also known as Mexican hat for the high-centered, broad-brimmed sombreros worn during fiestas. Indigenous tribes used a tea brewed from the leaves to treat rattlesnake bites. Prairie coneflower is a good pollinator-friendly and deer-resistant addition to home gardens.

DAGGERPOD
Phoenicaulis cheiranthoides
Mustard family (Brassicaceae)

Description: Perennial, often with sprawling reddish-purple flowering unbranched stems up to 12" long. Basal leaves in clusters are grayish white, elliptical to inversely lance-shaped, and 1"–7" long. Flowering stalks are up to 8" long and bear reddish-purple, pink, or white flowers. Flowers have 4 petals and are borne in dense clusters. Fruit is a daggerlike pod that can reach 4" long.

Bloom season: Mid-spring to early summer

Range/habitat: Western United States in open, dry areas such as grasslands or sagebrush flats, often in rocky, volcanic soils

Comments: *Phoenicaulis* is from *phoni* ("reddish purple") and *caulis* ("stem"), in reference to the color of the stems. *Cheiranthoides* ("hand of flowers") refers to the flowering stems spreading in different directions from the center. The daggerlike shape of the seedpods gives the plant its common name.

CRAIG MARTIN

SCARLET HEDGEHOG CACTUS

Echinocereus coccineus
Cactus family (Cactaceae)

Description: Perennial, low-growing cactus with few to many (over 100) clustered stems. Individual stems are cylindrical, may be up to 16" tall and 2" thick. Stems very from spineless to densely covered with spines; central spines are straight or slightly curved, and the radial spines are smaller. Scarlet to orange-red flowers have numerous petals and stamens, are about 3½" long, and are more tubular than dish-shaped. Fruit is a fleshy pod.

Bloom season: Spring to early summer

Range/habitat: Southwestern United States and northern Mexico in desert grasslands, shrublands, woodlands, and rocky areas

Comments: *Echinocereus* is from the Greek *echinos* ("hedgehog") and refers to the plant's resemblance to that animal. *Coccineus* ("orange red") refers to the color of the flowers. Hummingbirds are attracted to these flowers, which is unique since the flower shape fits the entire head, not just the bird's bill. A similar species, claret cup cactus (*E. triglochidiatus*), also has red flowers.

PATRICK ALEXANDER

ARIZONA RAINBOW CACTUS

Echinocereus rigidissmus
Cactus family (Cactaceae)

Description: Perennial, single-stem and barrel-shaped cactus, about 12" tall. Short reddish-purple or light-colored radial, comblike spines arise from the 15–23 ribs. Bright pink flowers are 2½"–4½" long, have a yellow throat or yellowish-white band at the base of the tepals, and have numerous stamens. Fleshy fruit is circular and dark green to purplish, with dark-colored seeds.

Bloom season: Spring to early summer

Range/habitat: Southwestern United States and northern Mexico in desert grasslands, desert slopes, and oak woodlands

Comments: *Echinocereus* is from the Greek *echinos* ("hedgehog") and refers to the plant's resemblance to that animal. *Rigidissmus* ("very rigid") refers to the spines. The common name refers to the different colors of the spines, which differentiate the new season's growth and are a rainbow of colors. *E. pectinata*, rainbow cactus, is similar but has a band of green below the inner white band.

STEVE R. TURNER

CARDINAL FLOWER

Lobelia cardinalis

Bellflower family (Campanulaceae)

Description: Perennial, grows 1'–6' tall. Lower, egg- to lance-shaped leaves are alternate, up to 8" long, with teeth along the margins. Upper leaves are 2" long and stalkless. The reddish, 2-lipped flowers are composed of 5 petals and borne along an elongated spike, often up to 28" long; the lower lip has 3 lobes and the upper lip has 2 lobes. Petals are united partway down to form a long tube. Flowers are up to 1¾" long. Fruit is a capsule.

Bloom season: Summer

Range/habitat: From northeastern Canada to the southeastern and southwestern United States in moist locations along streams, springs, sloughs, wetlands, and moist woodlands; also extends southward into Central America

Comments: *Lobelia* honors Matthias de l'Obel (1538–1616), Flemish physician and botanist, who devised a new (at the time) plant-classification system based on leaf structure and was one of the first botanists to understand the difference between monocots and dicots. *Cardinalis* ("cardinal red") refers to the scarlet color of the flowers, which resembles a Roman Catholic cardinal's red robes. The tubular flowers attract hummingbirds and butterflies as their primary pollinators.

ELEANOR DIETRICH

CLASPING-LEAVED VENUS'S LOOKING GLASS

Triodanis perfoliata

Bellflower family (Campanulaceae)

Description: Unbranched annual, 4"–30" tall. Stem and leaves bleed a milky sap when crushed. Stem is grooved, with rows of fine hairs growing along the edges. Lower leaves are alternate, stalkless, egg- to heart-shaped, and clasp the stem. A pair of opposite leaves at the top of the stalk are heart-shaped, about 1" long, with wavy or toothed margins. Arising from the upper leaf axils are 1–3 flower buds that open to ½"-wide violet to purple flowers with short tubular throats and 5 lobes. Flowers appear wheel-shaped, and the center of the throat is white. Fruit is a capsule with tiny seeds.

Bloom season: Late spring to midsummer

Range/habitat: North and South America in prairies, savannas, woodland edges, and disturbed areas

Comments: *Triodanis* is from the Greek *treis* ("three") and *odous* ("teeth"), referring to the 3-lobed calyx present on non-opening flowers. *Perfoliata* ("perfoliate") refers to the leaves clasping the stem. The seeds are shaken loose from the mature capsule by the wind and blown away from the parent plant. Some of the lower flowers are self-pollinating and never open. The common name refers to the shiny seeds of a related plant.

JIM FOWLER

FIRE PINK

Silene virginica

Pink family (Caryophyllaceae)

Description: Perennial, clump-forming; stems 12"–20" tall. Basal leaves are spatula- to inversely lance-shaped and up to 4" long and ¾" wide. Upper stem leaves are oppositely arranged, lance-shaped, and generally up to 6" long, but some may reach 12" long. Leaves and calyxes are covered with sticky, gland-tipped hairs. Trumpet-shaped, reddish flowers are 2" across; the 5 petals are deeply notched at the tip, flare outward, and are fused together to form a long corolla tube.

Bloom season: Mid-spring to midsummer

Range/habitat: Eastern North America in rocky woodlands, meadows, open woods, and thickets

Comments: *Silene* is for a Greek woodland deity, Silenus. *Virginica* ("of Virginia") refers to the eastern distribution of the plant. The common name refers to the scarlet flower color. Fire pink and royal catchfly (*S. regia*), which grows in the central United States, are both used in cultivated gardens and attract ruby-throated hummingbirds and butterflies as pollinators.

ROCKY MOUNTAIN BEEPLANT

Peritoma serrulata

Cleome family (Cleomaceae)

Description: Annual, 1'–6' or taller, often in clusters. Stems generally have numerous branches bearing compound leaves that are smooth to slightly hairy and have 3 lance-shaped to elliptical leaflets. Leaflets are ¾"–3½" long. Flowering stalks bear ½"- to 1"-long, pink to reddish-purple (sometimes white) flowers that have 4 petals and 7 long-exserted stamens. Slender seedpods are about 2" long and bear round, rough-textured seeds.

Bloom season: Summer

Range/habitat: Western and central United States in disturbed sites, sagebrush steppe, and dry flats

Comments: *Peritoma* is from the Greek *peri* ("cut around") and *tome* ("division"), referring to the cut around the calyx base. *Serrulata* ("finely saw-toothed") refers to the leaf edges. The flowers produce great quantities of nectar that attract bees; hence the common name. Native Americans in the Southwest ate the cooked leaves and also created black pottery paint from the boiled plants.

STEVE R. TURNER

WIDOW'S CROSS

Sedum pulchellum
Stonecrop family (Crassulaceae)

Description: Annual to perennial, depending on location, with stems 4"–12" tall. Stems tend to branch outward in a starlike pattern and may form a dense mat over a rocky surface. Stems and leaves are light green in color. Alternate leaves are densely arranged along the stem, up to 1" long, and cylindrical to linear in shape. Leaves lack a petiole. Fruit is a ¼"-long seedpod.

Bloom season: Late spring to midsummer

Range/habitat: South-central and southeastern United States, growing over rocky outcrops

Comments: *Sedum* comes from a word meaning "to sit," in reference to its low-growing habit over rocky areas. *Pulchellum* ("beautiful") refers to the flower color and form. The common name is in reference to the petals arranged in a cross pattern. Also known as limestone or rock stonecrop or rose stonecrop.

DAVID LEGROS

TRAILING ARBUTUS

Epigaea repens
Heather family (Ericaceae)

Description: Mat-forming perennial; stems 4"–6" tall but may extend up to 16" long. Stems are hairy. Leathery, evergreen leaves are oval shaped, alternately arranged, and ¾"–3" long, bearing rust-colored hairs below. Fragrant pink to whitish, shallow, tubular flowers are about ½" wide and have 5 fused petals forming a short tube that flares open and bears numerous hairs inside the throat. Fruit is a fleshy, berrylike capsule.

Bloom season: Spring

Range/habitat: Eastern North America in bogs, savannas, hillsides, and woodlands

Comments: *Epigaea* ("on the earth") refers to the low-growing habit of this plant. *Repens* ("creeping") also refers to this low growth habit. Also known as Plymouth Mayflower, for this was one of the early flowers to bloom after the *Mayflower* pilgrims' first winter in America. John Greenleaf Whittier's poem "The Mayflowers" highlights this contrast of the *Mayflower*'s white sails with the first flowers of spring. This plant is the floral emblem of Nova Scotia and the state flower of Massachusetts. Many Indigenous tribes in the plant's range used it medicinally to treat kidney ailments, abdominal pain, arthritis, and rheumatism.

ALAN ROCKEFELLER

GNOME PLANT

Hemitomes congestum
Heather family (Ericaceae)

Description: Perennial, ¾"–4" tall and fleshy. Scalelike leaves arise along the pink to cream flowering stalk that arises from the top of the rhizome. Flowers are tightly bunched together, cream to pinkish, and hairy on the insides. The 4 (sometimes 5) petals flare open at the tip and are fused together to form a tubular base; a stout yellow stigma occupies the center of the flower. Fruit is a fleshy white berry.

Bloom season: Late spring to midsummer

Range/habitat: Western United States in moist woods, especially in coniferous forests

Comments: *Hemitomes* ("half a eunuch") refers to one anther not having pollen. *Congestum* ("crowded") refers to the dense cluster of flowers. The common name describes the stature of the plant. Small mammals eat the fruit, which has a musky or cheesy aroma. The animal droppings then contain seeds, which pass through the animal's digestive system and help spread the plant. This is a difficult wildflower to locate due to its small stature and its growing in leaf litter in dense woods. The sticky nature of the clumped pollen grains helps the pollen to adhere to insects, and the inner hairs may prevent smaller insects from robbing the flower of nectar. However, which insects pollinate the flowers is a mystery. The plant lacks chlorophyll and is considered a (mycotroph) heterotroph, obtaining nutrients through association with mycorrhizal fungi.

STEVE R. TURNER

PINK SHINLEAF

Pyrola asarifolia
Heather family (Ericaceae)

Description: Perennial, 2"–14" tall. Basal cluster of leaves is glossy green, evergreen, long stalked, and rounded. The upright, reddish floral stem bears numerous pink, cup-shaped flowers that have waxy petals. Fruit is a capsule.

Bloom season: Summer

Range/habitat: Across much of North America in woodlands

Comments: *Pyrola* ("pear-like") refer to the shape of the leaves. *Asarifolia* ("*Asarum*-like leaves") refers to the similarity of the leaves to those of wild ginger (*Asarum*). The undersides of the leaves may be pink; hence the common name.

MAREK BOROWIEC

SNOW PLANT

Sarcodes sanguinea
Heather family (Ericaceae)

Description: Perennial, 6"–12" tall. Unmistakable red stalk bears densely arranged red flowers and numerous scalelike leaves that overlap below and curl away from the stem higher up. Tubular-shaped flowers flare open at the mouth and often hang downward. Fruit is a pinkish capsule.

Bloom season: Early spring to midsummer

Range/habitat: Western North America in coniferous forests

Comments: *Sarcodes* ("fleshy") refers to the texture of the stem and flowers. *Sanguinea* ("blood red") refers to the brilliant red color of the stem and flowers, which often appear as the snowline recedes. Like many other non-chlorophyll-bearing plants, snow plant is considered a mycotroph or monotropid, a plant that obtains nutrients through association with mycorrhizal fungi that are attached to tree roots. The plant provides fixed carbon to the fungi, so they mutually benefit.

SCARLET LOCOWEED

Astragalus coccineus
Pea family (Fabaceae)

Description: Low-growing perennial, up to 6" tall. Pinnately compound leaves are up to 4" long, with 7–15 oblong to inversely lance-shaped hairy leaflets with pointed tips. Leaves and flowers arise directly from the root crown. Reddish, pea-shaped flowers are borne in clusters of up to 10 flowers; flower stalks and calyx are covered with stiff hairs. Flowers are about 1½" long; the upper, or banner, petal bends backward about 30 degrees and has a white patch with red veins. The lower, or keel, petal projects forward. Fruit is a curved seedpod with dense hairs.

Bloom season: Spring to early summer

Range/habitat: Southwestern United States (California, Nevada, and Arizona) in gravelly sites in sagebrush, pinyon-juniper woodlands, and desert areas

Comments: *Astragalus* may be from the Greek *astragulos* ("anklebone"), referring to the shape of the seedpod. *Coccineus* ("scarlet") describes the brilliant color of these flowers. This species is the only red-flowered *Astragalus* in the United States. Livestock that consumes too much of this plant's foliage may go "loco."

KEN CHEEKS

SPURRED BUTTERFLY PEA

Centrosema virginianum
Pea family (Fabaceae)

Description: Perennial vine, 2'–12' long, without tendrils. Weak stems either lie prostrate or twine around other vegetation. Pinnately compound leaves are alternate, 1"–4" long, with 3 lance- to egg-shaped leaflets that are ⅓"–1" long. The long, pointed calyx lobes are equal to or exceed the length of the flower tube. Purplish to lavender flowers arise in a small calyx lobe set above the 2 lower lobes. Center of the flower has a white spot with purple nectar lines. Fruit is a hairy pod.

Bloom season: Spring and summer, but may bloom throughout the year

Range/habitat: Eastern United States south into South America in sandy fields, coastal areas, and woodlands

Comments: *Centrosema* is from *kentron* ("spur") and *sema* ("sign"). *Virginianum* ("of Virginia") describes where the plant was first collected for science. Atlantic pigeonwings (*Clitoria mariana*) is similar, but the pigeonwings' largest petal is scooped.

E. CHRISTINA BUTLER

EASTERN MILKPEA

Galactia regularis
Pea family (Fabaceae)

Description: Perennial vine, with stems up to 4' long that are prostrate or twine up surrounding bushes. Alternate compound leaves have 3 egg-shaped to elliptical leaflets, 1½"–2½" long and about half as wide. Stems are covered with fine hairs that give stems a downy or milky appearance. Pea-shaped flowers are violet-purple and about ½" long. Fruit is a seedpod.

Bloom season: Mid- to late summer

Range/habitat: South-central and southeastern United States in dry prairies, pinelands, sandhills, and sandy woods

Comments: *Galactia* is from the Greek *gala* ("milk-yielding"), referring to the misconception that the plant contains milky sap. English naturalist Patrick Browne (1720–1790) first named this plant. *Regularis* ("regular") refers to the common pealike appearance of this plant, which Linnaeus had first placed in the *Dolichos* genus. Also known as downy milkpea. Florida hammock milkpea (*G. striata*) has white striations on the upper petal and grows in Florida and Louisiana south into South America.

TOM LEBSACK

CATCLAW SENSITIVE BRIAR

Mimosa nuttallii

Pea Family (Fabaceae)

Description: Perennial, 1'–2' tall. Sprawling stems are 1'–6' long and bear recurved prickles. The doubly pinnately compound leaves are 2"–6" long and divided into 4–8 segment pairs. Each 1"- to 2"-long segment bears 8–15 pairs of ⅓"-long elliptical leaflets. Round flowering heads arise on 1"- to 3"-long stalks, are ¾"–1" wide, and bear numerous small 5-petaled pink to lavender flowers. Fruit is a slender, 1"- to 5"-long prickly seedpod.

Bloom season: Late spring through summer

Range/habitat: Central United States in dry prairies, woodlands, disturbed areas, and grasslands

Comments: *Mimosa* is from the Greek *mimos* ("to mimic"), referring to the leaves closing when disturbed, a movement that mimics the response of certain creatures. This action is called thigmonasty or seismonasty. *Nuttallii* honors Thomas Nuttall (1786–1859), an English botanist who collected plants and animals in the United States. The roots give off a foul odor when disturbed. The exserted stamens with yellow anthers obscure the petals. This plant is an important browse plant for livestock. Also known as Nuttall's sensitive briar or shame-boy for its leaves that close up when disturbed.

CHUCK TAGUE

SEASIDE GENTIAN

Eustoma exaltatum

Gentian family (Gentianaceae)

Description: Annual, but may be perennial in part of its range; 1'–2' tall with smooth stems. Oval-shaped leaves are slightly succulent, 1"–2½" long, oppositely arranged, with pointed tips. Bell-shaped flowers are 2"–4" wide with 5 (occasionally 7) bluish-purple petals that fuse together near the base to form a short tube. Flower throat has dark purplish markings surrounded by a white halo that contrasts with the green stigma and stamens. Fruit is a capsule.

Bloom season: Winter through summer

Range/habitat: Southern United States from Florida to California and north through the central states to Montana and south into Central America and the West Indies in sandy coastal areas, moist plains, and freshwater marshes

Comments: *Eustoma* is from the Greek *eu* ("good") and *stoma* ("mouth"), referring to the large open throat of the tulip-like flower. *Exaltatum* ("very tall") refers to the upright growth of the plants. The sap of the plant may cause a skin rash, but the plant has been used to treat eye ailments. Also known as catchfly prairie gentian, bluebells, or prairie gentian.

JIM FOWLER

ROSE PINK

Sabatia angularis
Gentian family (Gentianaceae)

Description: Annual or biennial, 1'–3' tall. From a basal rosette of ruffled-looking leaves arise multiple 4-angled, winged stems that branch in the upper portion. Upper leaves are opposite, stemless, lance- to heart-shaped, and up to 1½" long. Red to pink star-shaped flowers are arranged in flat-topped clusters, each flower about 1"–1½" wide with 5 petals that are joined at the base. Center of the flower has a starlike pattern of yellow to yellowish green outlined in dark rose. Fruit is a ⅓"-long capsule.

Bloom season: Summer

Range/habitat: Eastern North America to New Mexico in moist sites such as fields, prairies, meadows, open woods, and marshes

Comments: *Sabatia* honors Liberato Sabbati (1714–1778), an Italian botanist, surgeon, and curator of the Botanical Garden in Rome. *Angularis* ("angular") refers to the 4-angled stems. The flowers may also be white or lack the prominent dark edging in the flower's center. Bartram's rose gentian (*S. decandra*) grows in shallow freshwater wetlands and has numerous (10–14) corolla lobes.

STEVE R. TURNER

STICKY GERANIUM

Geranium viscosissimum
Geranium family (Geraniaceae)

Description: Perennial, 1'–3' tall, with sticky hairs on the stems and leaves. The highly dissected leaves are mostly basal. Borne on long stalks, the palmately compound leaves are opposite, up to 5" long, and divided into 5–7 lobes, which are again divided into segments that are cleft or toothed. Flowers vary from light pink to a dark pinkish purple and are about 1½" wide, saucer-shaped, with purple nectar lines. Fruit is a 2"-long sticky capsule.

Bloom season: Mid-spring through summer

Range/habitat: Western North America in meadows, stream banks, roadsides, and woodland forests; often in nutrient-poor soils

Comments: *Geranium* is from the Greek *geranos* ("crane"), referring to the seed capsule, which resembles a crane's bill. *Viscosissimum* ("sticky") refers to the glandular hairs on the flower bracts and leaves. Insects that become stuck to the sticky hairs of this protocarnivorous plant are dissolved by digestive enzymes produced by the plant in order to obtain nitrogen from the insect proteins. The flowers are edible, and Northwest Native Americans used the roots and leaves to treat colds, sore eyes, and stomach ailments.

STEVE R. TURNER

VIRGINIA WATERLEAF

Hydrophyllum virginianum
Waterleaf family (Hydrophyllaceae)

Description: Perennial, 1'–2' tall. If present, hairs on the stems and leaves are curved or appressed to the surface. Long-stalked leaves are 6"–10" long, alternate, deeply divided into 3–7 lobes or leaflets, with toothed edges and pointed tips. The first leaves have white spots that make them appear mottled; older leaves lack these spots. Two-inch-wide clusters of 8–20 small white to lilac, hairy flowers are borne on long stalks arising from leaf axils. Bell-shaped flowers have stamens that project far beyond the petal rims; the corolla lobes barely spread apart. Fruit is a capsule.

Bloom season: Mid-spring to early summer

Range/habitat: Eastern North America in river valleys, hillsides, and open woods

Comments: *Hydrophyllum* is from the Greek words *hydro* ("water") and *phyllon* ("leaf"), referring to the mottled leaves appearing water stained. *Virginianum* ("of Virginia") refers to part of the range for this plant. Certain years these plants may flower for a second time in late summer. Medicinally, a tea was made to treat dysentery and to stop bleeding; the pulverized roots were used as a balm to treat cracked lips.

TOM LEBSACK

COPPER IRIS

Iris fulva
Iris family (Iridaceae)

Description: Perennial, 2'–3' tall. Narrow, blade-shaped leaves are bright green, 2'–3' long, and bend away from the flowering stem with age. Stem leaves are smaller. Copper-colored flowers are up to 3" wide, have 6 widely spreading petal-like appendages, and are beardless and crestless. Flowers have 3 drooping sepals and 3 drooping petals; the petallike sepals are a little larger than the petals. Base of the flower is yellowish and tubular. Fruit is a seed capsule.

Bloom season: Mid-spring to early summer

Range/habitat: Southern and central United States in sloughs, ponds, bald cypress swamps, and other wetland areas that are spring flooded but dry in summer

Comments: *Iris* is for the Greek goddess Iris, goddess of the rainbow. *Fulva* ("tawny-orange") refers to the flower color, which may range from bronze to red and rarely yellow. The flowers are slightly fragrant. Also called red iris.

GRASS-WIDOW

Olsynium douglasii
Iris family (Iridaceae)

Description: Perennial, 4"–16" tall. Plants often bloom in profusion. Alternately arranged, 2 grasslike leaves are 4"–12" long. The flowering stalk bears 1 to several magenta flowers bearing 6 tepals and 6 stamens fused together for about half their length. Flowers are about 1½" wide with white centers and are bowl- to saucer-shaped. Fruit is a capsule.

Bloom season: Late winter to late spring

Range/habitat: Western North America in moist meadows, rocky areas, meadows, and open woods

Comments: *Olsynium* is from several Greek words meaning "hardly united," which refers to the stamens being partially united about halfway down the filaments. *Douglasii* honors David Douglas (1799–1834), a Scottish plant collector who traveled and collected extensively in the Pacific Northwest for the Royal Horticultural Society of London. Douglas met a tragic end, dying in Hawaii while on a collecting exploration. These plants may carpet large areas with spectacular blooms.

MARGARET MARTIN

NETTLE-LEAF GIANT HYSSOP

Agastache urticifolia
Mint family (Lamiaceae)

Description: Perennial; plants may be 3'–6' tall, with multiple square stems. Stems and leaves have a strong minty or anise-like odor when crushed. Leaves are broadly lance-shaped to triangular, opposite, 1"–3" long, light green below, with toothed edges. Flower head, 1"–7" long, is a dense cluster of small, violet to white flowers with 2 upper lobes and 3 lower ones. Petals are very short, and stamens extend beyond the petals. Fruit is a brown nutlet.

Bloom season: Summer

Range/habitat: Western North America in meadows, dry slopes, and brushy habitats

Comments: *Agastache* ("many spikes") refers to the numerous flowering stems. *Urticifolia* ("nettle leaved") refers to the leaves resembling those of stinging nettle (*Urtica dioica*), but they lack the stinging hairs of that plant. Dried leaves and flowers are brewed as an herbal tea, and the plant is used medicinally to treat rheumatism, stomach upsets, and colds. This widely distributed plant attracts numerous pollinators, including bees, hummingbirds, and butterflies such as the monarch butterfly. Also called horsemint.

KATIE BYERLY

WILD BERGAMOT

Monarda fistulosa
Mint family (Lamiaceae)

Description: Clump-forming perennial, with square stems 2'–5' tall. Upper portion of stems may be hairy. Oblong leaves are up to 5" long, opposite, hairless, and toothed along the edges. Rounded clusters of pink, red, lavender, or white flowers are borne on the ends of flowering stalks. The 2-lipped tubular flowers are about 1" long. Upper lip is narrow and hairy; lower lip is broad and 3-lobed. Each flower head is subtended by a whorl of large pinkish bracts. Fruit is a nutlet.

Bloom season: Summer

Range/habitat: Across most of North America in dry sites such as prairies, savannas, fields, woodlands, and roadsides

Comments: *Monarda* honors Nicolás Bautista Monardes (1493–1588), a sixteenth-century Spanish botanist and physician who was able to study the flora of the Americas due to Spanish explorers in the region. *Fistulosa* ("tubular" or "hollow like a pipe") refers to the flower shape. The leaves have an oregano-like aroma when crushed and were, and still are, used to treat respiratory illnesses. Also known as bee balm.

KATIE BYERLY

OBEDIENT PLANT

Physostegia virginiana
Mint family (Lamiaceae)

Description: Perennial, 1'–4' tall. Square stems bear opposite, narrowly lance-shaped leaves that are 3"–6" long, mostly stalkless, and sharply toothed along the margins. The 2-lipped flowers are white to rose to pale pink, have 4 stamens, and arise along an elongated stalk, maturing from bottom to top. Small leaflike bracts subtend the flowers. Fruit is a small seed.

Bloom season: Summer to fall

Range/habitat: Eastern half of North America from New Mexico to Quebec in open meadows, stream banks, prairies, and wooded areas

Comments: *Physostegia* comes from the Greek words *physa* ("bladder") and *stege* ("covering"), referring to the inflated calyx as it develops. *Virginiana* ("of Virginia") is in reference to the plant's distribution. The common name refers to the flowers temporarily staying in position after being rotated along the flowering stem.

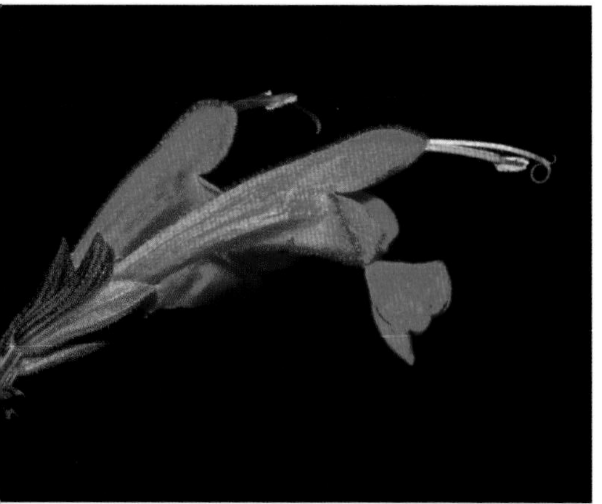

CHUCK TAGUE

TROPICAL SAGE

Salvia coccinea
Mint family (Lamiaceae)

Description: Perennial, 1'–4' tall, spreading outward to about 2½' wide with square stems. Variable leaves are up to 3" long, hairy, egg-shaped to triangular, and toothed along the margins. Borne on 1'-long spikes, the flowers are arranged in whorls around the flowering stalk. Tubular red flowers are 2-lipped, about 1¼" long, and have long exserted stamens. Fruit is a capsule with tiny seeds.

Bloom season: Spring and summer

Range/habitat: Southeastern United States to South America and the Caribbean in sandy forested areas, disturbed sites, and coastal areas

Comments: *Salvia* ("to save" or "to heal") refers to the medicinal properties of some species in this genus. *Coccinea* ("scarlet dyed") refers to the flower color. Though the plant resembles *S. elegans*, the leaves of tropical sage are not edible. In certain parts of its range, the plant may be an annual. This is the only native *Salvia* in the United States with red flowers. Also called Texas sage.

KEN KNEIDEL

TEXAS BETONY

Stachys coccinea
Mint family (Lamiaceae)

Description: Perennial, 1'–3' tall. Square stems with soft hairs bear oppositely arranged leaves that are triangular or egg-shaped and have toothed margins. Leaf size varies, ¾"–2½" long. The 2-lipped scarlet (sometimes orangish) flowers arise from leaf nodes along the stem, often in whorls of up to 6 flowers. The green calyx is sharply pointed and subtends the tubular flower, which is about 1" long and has a large, hoodlike upper lip. Lower lip curves downward and has 3 lobes. The 4 stamens have reddish stems and dark purple anthers. Fruit is a capsule.

Bloom season: Spring and summer

Range/habitat: Southwestern United States and south into Central America in moist shady locations along stream banks, canyons, and moist crevices.

Comments: *Stachys* ("ear of grain") refers to the elongated spike of flowers. *Coccinea* ("scarlet") refers to the flower color. This is the only member of this genus with red flowers. Hummingbirds are attracted to the flowers. Also called scarlet hedge-nettle for the leaves, which resemble those of stinging nettle (*Urtica dioica*).

STEVE R. TURNER

HAIRY HEDGE-NETTLE

Stachys pilosa
Mint family (Lamiaceae)

Description: Perennial, 1½'–3' tall. Square stems are light green to purplish in color and covered with fine hairs. Lance- to egg-shaped leaves are up to 4" long, saw-toothed along the margins, covered with fine hairs, and oppositely arranged. Flowers are arranged in whorls along an elongated stem that is 4"–8" long. Individual 2-lipped, tubular flowers are about ½" long, white to light pink, with rosy-white spots on the lips. The upper lip is hood-like and unlobed; the lower lip has 3 lobes. Fruit is a capsule with 4 nutlets.

Bloom season: Summer

Range/habitat: Across much of North America except the Southeast in moist locations along streams, wooded areas, marsh edges, roadsides, and prairies

Comments: *Stachys* ("ear of grain") refers to the elongated flower spike. *Pilosa* ("hairy") refers to the plant's long, soft hairs. This is the most wide-ranging *Stachys* in North America. Various types of bees are common pollinators of these flowers.

DAVID LEGROS

WOOD LILY

Lilium philadelphicum
Lily family (Liliaceae)

Description: Perennial, 1'–3' tall. Leaves are arranged alternately below and in a whorled pattern above, with 4–11 segments. Individual leaves are linear to elliptical and 1½"–4" long. The reddish-orange to orange (rarely yellow) flowers are funnel-shaped, 2"–3" long, and usually arise in clusters of 1–3 at the end of the flowering stalk. The 6 spatula-shaped tepals are narrow at the base, with yellowish patches spotted with purplish brown dots. Fruit is a 3-part capsule.

Bloom season: Summer

Range/habitat: Widespread across much of North America except the western United States in dry woods, prairies, and meadows

Comments: *Lilium* is from the Greek *lirion* ("a lily"). *Philadelphicum* ("of Philadelphia") refers to the type locality. The wood lily is the floral emblem of Canada's Saskatchewan province.

ELEANOR DIETRICH

INDIAN PINK

Spigelia marilandica
Logania family (Loganiaceae)

Description: Perennial, 1'–2¼' tall, clump forming. Glossy egg- to lance-shaped leaves are up to 4" long and borne in opposite pairs along the stem. Leaves lack stems. Trumpet-shaped scarlet to reddish flowers are borne upright along one side of a flowering stalk, up to 2" long, and flare open at the mouth. Inside of the flower is yellow, and the style rises above the flared petal tips. Fruit is a capsule.

Bloom season: Late spring to early summer

Range/habitat: Southeastern United States from New Jersey to Texas in moist woods, along stream banks, and the edges of wetlands

Comments: *Spigelia* honors Adriaan van den Spiegel (1578–1625), a Flemish professor of anatomy at Padua in northern Italy. *Marilandica* ("of Maryland") describes the plant's type locality. When the capsule is mature, it splits open and explosively casts the seeds out. Also known as woodland pinkroot and worm-grass because the roots were used historically to rid the intestine of parasites.

STEVE R. TURNER

PURPLE POPPY-MALLOW

Callirhoe involucrata
Mallow family (Malvaceae)

Description: Mat-forming perennial, 6"–9" tall, but spreading 3' or more. Some of the stems are vine-like. Deeply palmately divided leaves are alternate, with 5–7 narrow lobes divided into smaller lobes and up to 4" across. Cup-shaped, open flowers are 1½"–2½" wide, and each of the 5 petals has a white spot at the base. The slightly hairy calyx has 5 pointed tips and is subtended by 3 leafy bracts. Stamens form a central column inside the flower. Fruit is a capsule.

Bloom season: Spring to fall

Range/habitat: From south-central Canada to the southern United States in dry fields, meadows, and waste places

Comments: *Callirhoe* is for Callirhoe, daughter of Achelous, the Greek river god. *Involucrata* ("involucre") refers to a ring of bracts that surrounds several flowers. Also known as wine cups.

BORREGOWILDFLOWERS.ORG

DESERT FIVE-SPOT

Eremalche rotundifolia
Mallow family (Malvaceae)

Description: Annual, 3"–24" tall; stems generally unbranched. Borne on long reddish stems, the round leaves are up to 2½" wide and have toothed edges. Leaves and stems are covered with bristly hairs; hairs are longer on the stems. The purplish-pink to lilac flowers are spherical in shape, with 5 overlapping petals that are lighter in color toward the base and have a dark purplish/red irregularly shaped spot at the base. Fruit is a capsule.

Bloom season: Spring

Range/habitat: Mostly southern California, Nevada, and parts of Utah in Mohave Desert scrublands, open areas, washes, and barren areas

Comments: *Eremalche* is Greek in origin and means "lonely place," referring to the plants in this genus growing in seemingly deserted and desert landscapes. *Rotundifolia* ("round leaf") describes the shape of the leaves, which move and track the sun for maximum exposure. The flowers close each night and open partially during the day to resemble a globe.

ELEANOR DIETRICH

SCARLET ROSE-MALLOW

Hibiscus coccineus
Mallow family (Malvaceae)

Description: Perennial, 3'–8' tall, with a woody base. Palmately divided leaves are 5"–6" wide; each leaflet is linear to lance-shaped and slightly toothed along the margins. Leaves are similar to those of marijuana or hemp (*Cannabis sativa*). Flowers are 3"–6" in diameter, 5-petaled, and scarlet red. Flowers arise in the upper leaf axils and have a stout center column bearing stamens and pistils. Fruit is a seedpod.

Bloom season: Summer

Range/habitat: Southeastern United States in wetlands, marshes, swamps, and stream edges

Comments: *Hibiscus* is the old Greek and Latin word for the plants. *Coccineus* ("scarlet") refers to the color of the flowers. Also called swamp mallow for where it grows or Texas star hibiscus, although the plant does not naturally grow in Texas.

STEVE R. TURNER

HALBERD-LEAVED ROSE MALLOW

Hibiscus laevis
Mallow family (Malvaceae)

Description: Perennial, 3'–6' tall, with smooth stems. Alternately arranged leaves have slightly toothed margins, are up to 6" long, and have 2 wide-spreading lobes at the base and 1 longer lobe that resemble the tip of a medieval weapon called a halberd. Some leaves may have 5 lobes resembling a maple leaf, and the lobes are usually triangular-shaped. Pink to white cup-shaped flowers arise from leaf axils along the entire plant and are 3" long and up to 6" wide, with 5 overlapping petals. The flower's throat is a darker maroon or purple and has a stout reproductive column that projects just beyond the flower's mouth. Fruit is a capsule with large flat, hairy seeds, which may be kidney-shaped.

Bloom season: Midsummer to fall

Range/habitat: Central and eastern North America in wet or moist soils near streams, lakes, sloughs, ponds, and swamps

Comments: *Hibiscus* ("mallow") is the old Greek and Latin name for mallows. *Laevis* ("smooth") refers to the texture of the leaves and stems. Though the plant produces flowers over a long period, a single flower is in bloom during the flowering season, and that flower lasts for just a single day. Related to okra, when the leaves or stem are crushed or cut, they exude a slimy sap. Also called smooth rose mallow.

STREAMBANK HOLLYHOCK

Iliamna rivularis
Mallow family (Malvaceae)

Description: Perennial, 3'–6' tall. Maplelike leaves are alternate, 4"–8" long, and palmately lobed into 3–7 rounded segments that are coarsely toothed and slightly hairy with fine starlike hairs. Rose to white, disk-shaped flowers are borne in a loose cluster; each flower has 5 egg-shaped petals that overlap. Fruit is a capsule.

Bloom season: Summer

Range/habitat: Western North America, mainly east of the Cascade Mountains, in moist meadows, riparian areas, sagebrush steppe, or higher-elevation forests

Comments: *Iliamna* derivation is uncertain; it might be named for Alaska's Iliamna Lake, although no *Iliamna* species grow there. *Rivularis* ("growing by streams") refers to the moist habitat type this plant grows in. The plants regenerate by seeds not rhizomes, and the long-lived seeds are viable for more than 100 years, which enables this shade-intolerant plant to respond to disturbances such as wildfires or intensive logging.

KAREN E. ORSO

OREGON CHECKERMALLOW

Sidalcea oregana
Mallow family (Malvaceae)

Description: Perennial, with flowering stems 1'–5' tall. Upright stems bear star-shaped hairs on the lower half and are smooth above. Leaves lobed to palmately compound. Bowl-shaped flowers are pinkish to rose, have 5 petals, and are borne along an elongated stalk. Fruit is a round capsule.

Bloom season: Summer

Range/habitat: Western United States in moist meadows, bogs, seeps, and other moist locations

Comments: *Sidalcea* is a combination of two other genera, *Sida* and *Alcea*. *Oregana* ("of Oregon") refers to the plant's distribution. This checkermallow may bear bisexual flowers (both male and female parts in the same flower) or just female flowers without anthers.

MARIANA YUSTE RAMIREZ

COMMON UNICORN PLANT

Proboscidea louisiana
Devil's Claw family (Martyniaceae)

Description: Low bushy annual, 1'–2' tall, with spreading stems up to 3' long. The large 5"-wide, long-stemmed, palmately lobed leaves are heart-shaped and may be up to 1' long. Leaves are covered with sticky glands to which soil may adhere. The 5-lobed, tubular flowers vary in color from purplish cream to lavender to a creamy yellow that is spotted with purple. Small clusters of flowers arise in leaf axils. Fruit is a curved pod up to 4" long that is fleshy at first but then dries and splits open into 2 curved halves; each half may be up to 1' long.

Bloom season: Late spring to early fall

Range/habitat: Widely distributed across the United States and Mexico in dry fields, pastures, and disturbed areas. The plant's native origin is thought to be the south-central United States but has expanded its range.

Comments: *Proboscidea* is from the Greek proboscis ("elephant's trunk"), which refers to the curved ends of the seedpods. *Louisiana* ("from Louisiana") refers to the type locality of the plant. The glandular hairs on the leaves and stems produce an oily, sticky secretion that has an acrid odor. When young, the seedpods are edible and are often soaked in a pickling brine. Native Americans and modern-day wild plant harvesters make a black dye from the pods. Also known as ram's-horn or devil's horn for the shape of the dried seedpods.

DAVID LEGROS

RED TRILLIUM

Trillium erectum
Bunchflower family (Melanthiaceae)

Description: Perennial, 6"–24" tall, with a single stem. Botanically speaking, the stem is an extension of the underground rhizome and the leaves are bracts that subtend the flowers. A whorl of 3 broad, diamond-shaped leaves are 3"–8" long and smooth on both sides. Above the leaves, a single maroon flower arises on a 1"- to 4"-long stem. The nodding flower is 2"–3"wide and has a foul aroma when open. Flower has 3 green sepals and 3 reddish petals, which recurve; petals are occasionally white, yellow, or pink. Fruit is a dark red, berrylike capsule.

Bloom season: Spring

Range/habitat: Eastern North America in moist woods

Comments: *Trillium* ("in threes") refers to the number of leaves and floral parts. *Erectum* ("erect") refers to the upright posture of the plant. A poultice from this plant has been used to treat gangrene and ulcers. Native Americans used the root to aid in childbirth. The foul-smelling flowers attract pollinators such as flesh flies and carrion beetles. The seeds have an oil-rich appendage on the tip that attracts ants, which carry the seeds to a nest and help disperse the plant. Also known as wet dog trillium, stinking Benjamin, or red wakerobin.

ELEANOR DIETRICH

MARYLAND MEADOW BEAUTY

Rhexia mariana
Melastome family (Melastomataceae)

Description: Perennial, 1'–2½' tall. Lance-shaped leaves are hairy, about 2½" long, arise in pairs, and are distinctively 3-veined. The 2"-wide flowers are rose to white or purple and arise in the leaf axils. The 4 petals are attached at the base to form a cylindrical tube, and the cluster of yellow or orangish stamens contrasts with the colorful petals. Fruit is an urn-shaped capsule.

Bloom season: Midsummer to fall

Range/habitat: Eastern and southern United States near water areas such as bogs, marshes, savannas, meadows, and disturbed sites

Comments: The derivation of the genus name *Rhexia* is a bit obscured; Dr. Asa Gray once wrote about the name being "applied to this genus without obvious reason." *Mariana* refers to the white drops of milk that spilled from the Virgin Mary when she nursed the baby Jesus. Bumblebees are the primary pollinators for these plants. When a bumblebee lands on a flower, it triggers the release of pollen. Thoreau once said that the urn-shaped capsule of meadow beauties reminded him of a little cream pitcher. Also known as pale meadow beauty.

JOHN POLITES

RED MAIDS

Calandrinia ciliata

Miner's Lettuce family (Montiaceae)

Description: Annual, 7"–14" tall; stems erect or spreading. Linear to lance-shaped leaves are semi-succulent, alternate, fringed with hairs along the margins, and ¾"–4" long. Bowl-shaped flowers are borne (in a raceme) on short stems and have 5 (3–7) pink to reddish petals. Fruit is a capsule.

Bloom season: Spring

Range/habitat: Western North America from British Columbia to New Mexico in deserts, grasslands, fields, and open forests; also occurs in Central and South America

Comments: *Calandrinia* honors Jean-Louis Calandrini (1703–1758), a Swedish professor of mathematics and philosophy but also a noted botanist. *Ciliata* ("fringed") refers to the fine hairs along the edges of the sepals. Native Americans collected and ate the seeds.

COLORADO FOUR O'CLOCK

Mirabilis multiflora

Four O'clock family (Nyctaginaceae)

Description: Perennial, clump-forming plant; 1'–3' tall, but may sprawl along the ground, covering a large area. Opposite, short-stemmed, glossy leaves are rounded to egg-shaped, ¾"–7" long and pointed at the tip. Flowers grow in small clusters in the leaf axils; individual magenta to lavender flowers are 1½"–2½" long and funnel-shaped. The broad 5 lobes are petallike sepals, and the pistil and stamens protrude beyond these sepals. The pointed bracts that subtend the flower are covered with glandular, sticky hairs that sand grains adhere to. Fruit is a capsule.

Bloom season: Late spring to early fall

Range/habitat: Southwestern United States from California to Texas and south into northern Mexico in desert foothills, canyons, or pinyon-juniper woodlands

Comments: *Mirabilis* ("marvelous") describes the overall appearance of the plant. *Multiflora* ("many flowered") refers to the abundance of flowers, which open in the afternoon and evening to attract pollinators such as butterflies, bees, hummingbirds, and hawkmoths. The Zuni were known to mix the powdered roots into flour for bread to decrease one's appetite. This plant was first collected for science along the Platte River in Colorado. Also known as showy four o'clock.

PATRICK ALEXANDER

SCARLET MUSK-FLOWER

Nyctaginia capitata
Four O'clock family (Nyctaginaceae)

Description: Perennial, 1'–3' tall, but sprawling; stems and leaves bear sticky hairs. Narrowly triangular-shaped leaves are ½"–3" long, gray-green, opposite, and have wavy to smooth margins. The 5-lobed, trumpet-shaped flowers are red to pink (rarely yellow) in color and are borne in dense clusters. The 1"- to 1½"-long flowers have a musky scent and sometimes have yellow streaks in the petals. The long, exserted stamens project far beyond the rim of the flower. Fruit is a capsule.

Bloom season: Spring to early fall; flowering usually associated with rainfall

Range/habitat: New Mexico and Texas and into northern Mexico in dry grasslands, roadsides, abandoned fields, rocky slopes, or shrublands

Comments: *Nyctaginia* is from the Greek words *nyctos* ("night") and *anthos* ("flower"), for the night-blooming characteristic of this plant. *Capitata* ("like a head") refers to the dense, rounded clusters of flowers. Hummingbirds pollinate the flowers; so do night-flying insects such as moths, which are also attracted by the musky aroma. Also known as devil's bouquet for the flower color and night-blooming habit.

FRANCISCO FARRIOLS SARABIA

TROPICAL ROYALBLUE WATERLILY

Nymphaea elegans
Waterlily family (Nymphaeaceae)

Description: Aquatic perennial. Rounded leaf blades, notched at the base, are 4"–10" in diameter and lie flat on the water surface. Greenish above, the blades are purple below. Solitary flowers, arising on fleshy stems 4"–6" above the water, are 2½"–5" wide and have 8–27 purple-tinged or white petals surrounding a center of numerous stamens. The green sepals and petals are arranged in multiple whorls. Fruit is a globe-shaped seed.

Bloom season: Spring to fall

Range/habitat: Southeastern United States to Colombia in ponds, lakes, sloughs, swamps, marshes, and ditches.

Comments: *Nymphaea* ("water nymph") refers to the naiads, or water nymphs, of Greek mythology, which were often found near streams or ponds. *Elegans* ("elegant") refers to the flowers. The petals are arranged in whorls of 4 and attract bees and beetles as their primary pollinators. The large leaves form platforms that rails and small herons may walk across.

FIREWEED

Chamerion angustifolium

Evening Primrose family (Onagraceae)

Description: Perennial, 2'–9' tall; stems often unbranched. The stalkless leaves are lance-shaped and 2"–10" long. Clusters of flowers arise at the ends of the stems. The rose to reddish-purple flowers are 1"–2" wide, with 4 petals and a 4-lobed stigma. The long, podlike capsule splits open to release numerous white-haired seeds.

Bloom season: Midsummer to fall

Range/habitat: Widely distributed across much of North America except the Southeast in clearings, disturbed woodlands, meadows, avalanche paths, and roadsides.

Comments: *Chamerion* is from the Greek *chamae* ("lowly") and *nerion* ("oleander") and refers to the plant's resemblance to a low-growing oleander. *Angustifolium* ("having narrow leaves") indicates the lance-shaped leaves. The common name describes the habit of this plant to colonize recently burned or disturbed areas; seeds may lie dormant for years until there is a disturbance. After the 1980 Mount St. Helens eruption, fireweed was one of the first plants to sprout in the area. The upright spikes of flowers attract bees as pollinators. During World War II, fireweed was known in Great Britain as bombweed for the plant's rapid colonizing of bomb craters.

DEERHORN CLARKIA

Clarkia pulchella

Evening Primrose family (Onagraceae)

Description: Annual, 4"–20" tall, but some plants may reach 3' tall. Linear leaves are arranged alternately along the stem. Lavender to rose-purple flowers have 4 petals that are more or less 3-lobed at the tip and narrow (clawed) at the base. Two sets of stamens may be present, with one set being fertile and the other infertile and smaller. Fruit is a slender capsule.

Bloom season: Late spring to early summer

Range/habitat: Portions of western and northern North America in dry grasslands, meadows, or open woodlands

Comments: Frederick Pursh (1774–1820) was a botanist who worked on the Lewis and Clark Expedition's collection, and he named the genus for one of the expedition leaders, William Clark (1770–1836). *Pulchella* ("pretty") refers to the lovely flowers. Meriwether Lewis, the other expedition leader, first recorded this plant for science on June 1, 1806. His description of the plant started: "I met with a singular plant today in blume of which I preserved a specemine." Also known as ragged robins, beautiful Clarkia, or elkhorns, for the shape of the petals.

HUMMINGBIRD-TRUMPET

Epilobium canum
Evening Primrose family (Onagraceae)

Description: Perennial with upright or mat-forming stems. Stems are hairy and may have sticky hairs along the margins. Reddish-orange flowers are 1"–1½" long and tubular-shaped, with tips that flare open. The 4 petals are 2-lobed at the tip, with stamens and a pistil extending beyond the mouth of the flower. Fruit is a capsule.

Bloom season: Midsummer to fall

Range/habitat: Western North America in rocky drainages, talus slopes, roadsides, and forest edges

Comments: *Epilobium* ("upon the pod") refers to the placement of the flower's ovary beneath the attachment of the petals. *Canum* ("gray" or "ashy") refers to the color of the leaves. The flowers attract butterflies and hummingbirds as pollinators; the plant is important to both groups, as the flowers provide nectar resources late in the summer season. Because of the drought-tolerant nature of the plant, it is often used in xeriscaping and roadside plantings. Also known as California fuchsia.

STEVE R. TURNER

SHOWY EVENING PRIMROSE

Oenothera speciosa
Evening Primrose family (Onagraceae)

Description: Perennial, 8"–24" tall, but with sprawling stems. Narrowly lance- to inversely lance-shaped leaves are 1"–3" long and may have short lobes near the base. Bowl-shaped flowers are 2"–3" wide and have 4 pinkish to white petals with fine rose-colored streaking. Flowers age to a darker shade of pink. Fruit is an oval capsule about 2" long.

Bloom season: Spring to midsummer

Range/habitat: Native to the southern half of the United States and northern Mexico in rocky meadows, plains, disturbed areas, and grasslands

Comments: *Oenothera* is from the Greek *oinos* and *theras* meaning "wine-seeker," for the roots of some species being used in making wine. *Speciosa* ("large") refers to the large colorful flowers that open late in the day and bloom throughout the evening, attracting moths and hawkmoths as pollinators. Also known as pink ladies, Mexican primrose, or pink buttercups, for the flower's resemblance to yellow flowers in the *Ranunculus* genus.

JIM FOWLER

DRAGON'S MOUTH ORCHID

Arethusa bulbosa
Orchid family (Orchidaceae)

Description: Perennial, 6"–15" tall. Plant has 1–3 grasslike leaves, which arise after flowering and may be up to 7" long. Solitary pink flower has 3 pinkish sepals, 1"–1¾" long, which arise behind the petals. Two of the petals are similar to the sepals, and these 2 form a hood over the lower lip. Lower lip (labellum) lacks a pouch and is whitish pink with purplish spots and a bristly yellow center. Fruit is a capsule.

Bloom season: Late spring through summer

Range/habitat: Eastern and central North America in bogs, swamps, and coniferous wetlands

Comments: *Arethusa* is for the Greek naiad, or nymph, Arethusa, who lived at a sacred spring on an island near Sicily. *Bulbosa* ("bulbous") refers to the small pseudobulb, or corm. Like many orchids, the flowers provide minimal or no nectar rewards for pollinators; bumblebees are attracted to the colorful flowers, which dust the insects with pollen as they back out from inside the flower's opening. The flowers depend on continued generations of inexperienced bees to provide pollination services. This is the only species in this genus.

STEVE R. TURNER

GRASS PINK ORCHID

Calopogon tuberosus
Orchid family (Orchidaceae)

Description: Perennial, up to 2½' tall. A single grasslike leaf is 3"–14" long. A cluster of 2–15 (rarely 20) pink (sometimes white) to rosy pink flowers with 3 petallike sepals and 3 petals are borne along an elongated stalk. Flowers are 1"–1½" wide, and the anvil-shaped labellum (a modified petal) is arranged at the top rather than the bottom like many other orchid species, with an inflated triangular tip. Center of the lip is a dense cluster of yellow stamen-like bristles. Fruit is a capsule.

Bloom season: Summer

Range/habitat: Eastern and central North America, as well as Cuba and the Bahamas, in sandy swales, bogs, swamps, wetland edges, and fens

Comments: *Calopogon* is from the Greek *kalos* ("beautiful") and *pogon* ("beard") and refers to the yellow, stamen-like hairs found on the flower's upper petal. *Tuberosus* ("tuberous") refers to the cylindrical underground corms. Grass pink is another orchid species with showy flowers that attract pollinators—in this case long-tongued bees such as carpenter bees, bumblebees, and leaf-cutter bees—but does not provide nectar or pollen rewards to the pollinators. Knobs on the ends of the hairs resemble pollen, which may attract the insects. When a bee visits the flower, pollen bags (pollinia) become snared on the upper side of the bee's abdomen, where the bee can't reach them; these are then transferred to another flower the insect visits in hopes of finding a reward. The common name refers to the grasslike leaves and the flower color.

FAIRY SLIPPER

Calypso bulbosa
Orchid family (Orchidaceae)

Description: Perennial, growing from a marble-size bulb. A single flowering stem arises 4"–8" from the bulb. The single oval-shaped leaf often fades during the summer. The 1"-long flower varies from pink to rose-purple (sometimes white or cream) in color. A single inflated petal forms the "slipper" portion of the flower and bears 2 hornlike projections at the base. This petal has purplish spots and streaks. Above this petal is the hood, which cloaks the stamens and styles. Above the hood are 2 more petals and 3 sepals, which are lance-shaped and project up and outward. Fruit is a capsule.

Bloom season: Late spring and early summer

Range/habitat: Found throughout the Northwest in coniferous forests from sea level to high elevations; occurs from Alaska throughout Canada and south to California, Colorado, and Arizona

Comments: *Calypso* is named for the sea nymph Kalypso, which is from a Greek word meaning "covered or hidden." *Bulbosa* ("bulbous") refers to the swollen underground stem. These orchids are pollinated by inexperienced bees, which seek the nectar rewards the plants advertise through fragrance and coloration but do not deliver. Mature bees leave the flowers alone, and the plants depend on successive generations of immature bees to continue pollination services. Also known as deer orchid.

JIM FOWLER

AUTUMN CORALROOT

Corallorhiza odontorhiza
Orchid family (Orchidaceae)

Description: Perennial, 5"–15" tall, from a rounded coral-like rhizome. Leaves are reduced to small sheaths that surround the base of the yellow-green to purple-brown flowering stalk. Reddish-brown to purplish-brown flowers are borne at the end of the flowering stalk and have a greenish cast. Both open (chasmogamous) and closed (cleistogamous) flowers are produced; the lip on the open flowers is often white with purplish spots. Fruit is a capsule.

Bloom season: Midsummer to fall

Range/habitat: Eastern North America from southern Ontario to Florida in humus-rich soils in coniferous or deciduous forests

Comments: *Corallorhiza* ("coral-like root") and *odontorhiza* ("toothed-root") both refer to the shape of the underground stems and rhizomes, which resemble sea coral, and the flowering stalk's toothlike swollen base.

DAVID LEGROS

PINK LADY'S SLIPPER

Cypripedium acaule
Orchid family (Orchidaceae)

Description: Perennial, 6"–15" tall. From near the base of the stem arise 2 oval-shaped leaves with deep venation; leaves are partially folded longitudinally and are 4"–8" long. The flowering stalk arises from between the leaves and bears a single, pouch-looking pink flower, 1½"–2½" long. A single greenish bract arches over the large lip, which is inflated and cleft on the upper side. The pouch may have reddish to greenish-brown veins. Fruit is a capsule.

Bloom season: Late spring to summer

Range/habitat: Across much of Canada and into parts of the central and eastern United States in moist or dry woods, bogs, sand dunes, and deciduous or coniferous forests

Comments: *Cypripedium* is from the Greek *cypris*, which refers to the goddess Aphrodite, and *pedilon* ("sandal"), for the fused petals that form a sandal-like pouch. *Acaule* ("without a stem") refers to the leaves, which lack a stem. This orchid is the provincial flower of Prince Edward Island, Canada, and the state flower of New Hampshire. The slipper-like lip of the flower gives these and some other species their common name. Many orchid seeds lack nutrients for the germinating seeds to survive on. The seeds require an association with a soil fungus in the *Rhizoctonia* genus to break open the seeds and supply them with water and food to survive. As the plant matures, it provides nutrients back to the host fungus. Also known as moccasin flower.

JIM FOWLER

ROSE POGONIA

Pogonia ophioglossoides
Orchid family (Orchidaceae)

Description: Perennial, 1'–3' tall. The flowering stalk bears a single leaf about midway up the stem, and the linear leaf is up to 4" long and 1½" wide. Generally, a single fragrant rose-pink flower opens at the top of the flowering stalk. Sometimes there may be 2 or 3 flowers, and occasionally the flowers are white. The lower lip's bearded throat is made up of numerous white or yellow hairs. The fruit is a capsule.

Bloom season: Mid-spring to midsummer

Range/habitat: Widespread across central and eastern North America in sphagnum bogs, fens, wet meadows, swamps, and wet woodlands

Comments: *Pogonia* is from the Greek word *pogon* ("haired" or "bearded") and refers to the appearance of the flower's lower lip, or labellum. *Ophioglossoides* is derived from several Greek words to mean "snake tongued," also in reference to the flower's lower lip. Plants are pollinated by bumblebees, which stick their heads into the flower in search of nectar or pollen. The top of the bee's head comes into contact with a stamen, which deposits pollinia (pollen bags) there, and the pollen is transferred to a stigma on the next flower the bee visits. The fragrant flowers have been described as smelling like raspberries, and in *Florida Wild Flowers* (1926), Mary Francis Baker wrote: "tantalizes with a suggestion of many perfumes." Also known as snake-mouth orchid and beard flower.

JIM FOWLER

PURPLE FRINGELESS ORCHID

Platanthera peramoena
Orchid family (Orchidaceae)

Description: Perennial, 1'–4' tall. Upright stems bear 2–5 lance- to narrowly egg-shaped leaves that are up to 8" long and 2" wide. The 3"- to 8"-long flowering head bears a cluster of bright rose-pink to purplish flowers; each flower is about 1" long and has 3 petallike sepals and 3 petals and a 1"-long nectar spur. The lower lip is deeply divided into 3 fan-shaped lobes, which are slightly toothed along the margins; the upper sepal and 2 petals form a hood. The middle lobe is notched. Fruit is a capsule.

Bloom season: Mid- to late summer

Range/habitat: Central and southeastern United States in swamps, marshes, roadsides, wet meadows, shrublands, and woodlands.

Comments: *Platanthera* ("broad anther") describes the shape of the anther. *Peramoena* ("very loving or beautiful") refers to the flowers. Butterflies and sphinx moths are attracted to the flowers as pollinators. The "fringeless" common name refers to the slight fringe on the edges of the petals.

DESERT PAINTBRUSH

Castilleja chromosa
Broomrape family (Orobanchaceae)

Description: Perennial, 4"–20" tall. Often grows with clustered stems that are covered with fine hairs. Lower leaves are linear to lance-shaped; upper leaves have 1–3 lobes. All have fine, stiff hairs and may have a purplish tinge to them. Flowerlike bracts are reddish, hairy, and cupped; the cleft calyx lobes have rounded segments. Corolla is green and 2-lipped, the upper lip beak-like and the lower lip shorter and 3-toothed. Fruit is a capsule.

Bloom season: Early to mid-spring

Range/habitat: Much of western North America in grasslands, sagebrush flats, and desert foothills

Comments: *Castilleja* is in honor of Domingo Castillejo (1774–1793), an eighteenth-century Spanish botanist and professor. His countryman José Celestino Mutis named the genus for Castillejo from plants Mutis collected in Colombia. *Chromosa* ("red") refers to the color of the flowers. Many paintbrushes are partially parasitic on the roots of other plants, specifically some in the Aster family, such as rabbitbrush, sagebrush, and snakeweed.

STEVE R. TURNER

PAINTED CUP

Castilleja coccinea

Broomrape family (Orobanchaceae)

Description: Biennial, 8"–24" tall, with generally unbranched, hairy stems. A basal rosette bears egg- to lance-shaped leaves; these leaves often wither before the flowers bloom. The 3"-long alternate stem leaves have 3–5 lobes and are hairy. Individual, ¾"-long flowers form dense clusters at the top of the plant. The colorful portion of the flowers is scarlet or orange (yellow forms also occur), 3-lobed bracts that surround the narrow greenish-yellow corolla with a short lip. Fruit is a capsule.

Bloom season: Mid-spring to midsummer

Range/habitat: Eastern to central United States and central Canada in prairies, thickets, fields, and moist or dry woodlands

Comments: *Castilleja* is in honor of Domingo Castillejo (1774–1793), an eighteenth-century Spanish botanist and professor in Cadiz, Spain. The genus contains over 200 species and was formerly in the Scrophulariaceae, or Figwort, family. *Coccinea* ("scarlet") refers to the flower color. This hemiparasitic plant, like other members of the genus, can survive on its own or may parasitize the roots of nearby plants—in this case perennial grasses or other wildflowers—to obtain nutrients. The primary pollinator for this plant is ruby-throated hummingbirds, although bees are also attracted to the flowers. Members of the Cherokee tribe made a toxic tonic from the plant to "poison their enemies"; members of the Menominee tribe placed parts of the plant onto another person as a love potion.

BORREGOWILDFLOWERS.ORG

GIANT RED PAINTBRUSH

Castilleja miniata

Broomrape family (Orobanchaceae)

Description: Perennial, 1½'–3' tall. Narrow lance-shaped leaves are 1"–2⅓" long, have entire edges, and are covered with thin hairs. Flowers are composed of colorful scarlet to red (sometimes pink) or orange bracts that surround a greenish-yellow, tubular corolla with red edges. The ½"- to 1¼"-long bracts are often deeply lobed but not always, and the middle lobe is much wider than the 2 outer lobes. Fruit is a capsule.

Bloom season: Late spring through summer

Range/habitat: Western North America and much of Canada in a wide range of moist habitats from sea level to mountain meadows

Comments: *Castilleja* is in honor of Domingo Castillejo (1774–1793), an eighteenth-century Spanish botanist and professor in Cadiz, Spain. *Miniata* ("colored red") refers to the flowers. These paintbrushes are hemiparasitic on other plants but may also produce their own nutrients through photosynthesis. The bright red flowers attract hummingbirds and bumblebees as pollinators. Also known as meadow paintbrush.

ELEPHANT'S HEAD

Pedicularis groenlandica
Broomrape family (Orobanchaceae)

Description: Perennial, 6"–20" tall. Leaves are fernlike, 2"–6" long, and toothed along the margins; when first emerging, the leaves may be more reddish. Reddish-purple, unbranched stems bear dense clusters of flowers at the terminal end. Each small, pinkish flower resembles an elephant's head. The upper petal is round, resembling the head, and then tapers to the long trunk. Three lower petals resemble the 2 large ears and the lower mouth. Fruit is a capsule.

Bloom season: Late spring to midsummer

Range/habitat: Western North America and Greenland in wet meadows and along stream banks at mid- to high elevations.

Comments: *Pedicularis* ("of lice") refers to the belief that herders or livestock grazing in fields of elephant's head would become infested with lice. The opposite interpretation is also possible—that the herb was a treatment for lice. *Groenlandica* ("of Greenland") refers to the type locality of this species near Nuuk, Greenland. Bumblebees pollinate these flowers by perching on the flower's trunk and forcing their way into the flower. The tongue-in-cheek definition of *Pedicularis* is a cross between "peculiar" and "ridiculous," for the bizarre shape of these flowers! Little elephant's head (*P. attollens*) is similar, but the flower's trunk barely extends beyond the throat.

STEVE R. TURNER

VIOLET WOOD SORREL

Oxalis violacea
Wood-Sorrel family (Oxalidaceae)

Description: Perennial, up to 16" tall. Basal leaves are long-stemmed (at first longer than the flowering stalks) and divided into 3 inversely heart-shaped leaflets that are bluish gray to greenish gray above and green or reddish below and resemble a shamrock. Leaflets are about 1" wide. Small clusters of bell-shaped pinkish-purple to lavender flowers (some are white) arise on weak stems, which tends to make them droopy. Each flower has 5 greenish sepals and 5 petals that are ½"–¾" long and fused at the base; the sepals may have an orangish or purplish spot at their tips. The flower's throat may be greenish white with fine purple streaks. Fruit is a capsule.

Bloom season: Mid-spring to early summer; may have a secondary bloom period later in the season

Range/habitat: Much of the central and eastern United States in moist prairies, fields, glades, along stream banks, and in woodlands

Comments: *Oxalis* is from a Greek word meaning "sour," which describes the taste of the leaves. *Violacea* ("violet") refers to the flower color. The Cherokee and Pawnee used this wood sorrel as a medicinal to treat various ailments. The tangy leaves and flowers are edible but contain oxalic acid. The habit of the flowers and leaves folding up at night or on overcast days is called nyctinasty.

BLEEDING HEART

Dicentra formosa

Poppy family (Papaveraceae)

Description: Perennial, spreading by underground roots. Stout stems are 8"–20" tall and bear basal compound leaves on long stems that are fern-like—highly dissected and lobed. The flowering stalk bears 4–15 pinkish (sometimes white), heart-shaped flowers that hang downward. Of the 4 petals, the outer 2 have short spurs that spread outward. Fruit is a podlike capsule containing black seeds.

Bloom season: Late spring and summer

Range/habitat: Western North America in moist woodlands and along stream banks

Comments: *Dicentra* is from the Greek *dis* ("twice") and *kentron* ("a spur"), for the 2 spurs formed by the petals. *Formosa* ("beautiful") describes the appearance of the flower. The seed tips have a tiny, white appendage that is rich in oil and is coveted by ants for food. The ants carry the seeds to their nest, acting as dispersal agents. Hummingbirds pollinate the flowers.

SCARLET MONKEY-FLOWER

Erythranthe cardinalis

Lopseed family (Phrymaceae)

Description: Perennial, 1'–3' tall; upright or with spreading stems. Downy leaves are opposite, oblong or inversely egg-shaped, palmately veined, pale yellowish green, and toothed along the margins. Red or reddish-orange flowers are about 1½" long and subtended by a hairy, green calyx. The 2-lipped flowers have an upper lip that arches forward and a lower lip that curves backward, exposing yellow hairs, the stamens, and style. Fruit is a capsule.

Bloom season: Late spring through summer

Range/habitat: Southwestern United States into southern Oregon and northwestern Mexico in moist areas such as seeps, springs, stream banks, roadside ditches, and wetlands.

Comments: *Erythranthe* is from the Greek *erythros* ("red") and *anthos* ("flowers"), referring to the red flowers. *Cardinalis* ("red") also describes the color of these flowers, which attract hummingbirds and butterflies as pollinators. David Douglas (1799–1834) was a Scottish botanist who collected plants in the western United States for the Royal Horticultural Society in London. His specimens sent back to London created an obsession for this species with European gardeners.

CHUCK TAGUE

LEWIS' MONKEY-FLOWER

Erythranthe lewisii

Lopseed family (Phrymaceae)

Description: Perennial, often growing in thick clusters. Stems are 15"–45" tall and covered with soft, sticky hairs. Oval-shaped leaves clasp the stem and are arranged oppositely. Leaves are toothed along the margins. Trumpet-shaped, rose-red or pinkish flowers are 1½"–2" long and 2-lipped. The lower lip has 2 yellowish, hairy ridges and 3 lobes; the upper lip has 2 lobes. Fruit is a capsule.

Bloom season: Summer

Range/habitat: Western North America in moist clearings, stream banks, and rocky seeps

Comments: *Erythranthe* is from the Greek *erythros* ("red") and *anthos* ("flowers"), which refers to the reddish flowers. *Lewisii* honors Meriwether Lewis (1774–1809), who, with Capt. William Clark, led the Corps of Discovery Expedition across the United States in 1804 under the guidance of President Thomas Jefferson. Lewis received botanical training prior to the expedition, and Lewis and Clark collected numerous plant and animal specimens that were new to science. The insides of the flowers have flypaper-like glands that trap insects; presumably the trapped insects decompose and provide nutrients for the plants.

ALLEGHENY MONKEY-FLOWER

Erythranthe ringens

Lopseed family (Phrymaceae)

Description: Perennial, 1'–3' tall (occasionally up to 4'), with square stems. Stalkless, opposite leaves clasp the stem and are oblong to lance-shaped, 2"–4" long, and may be sharply toothed along the margins. The 2-lipped flowers are bluish purple (sometimes white or pink), 1" long, and have a ½"-long calyx tube. Flowers arise in pairs from leaf axils along the upper stem on stalks that are longer than the calyx tube. The upper lip of the snapdragon-like flower is divided into 2 erect lobes; the lower lip is divided into 3 rounded lobes with pale white or yellowish spots and yellow hairs at the base of the middle lobe. Fruit is a rounded capsule.

Bloom season: Summer to early fall

Range/habitat: Occurs across much of North America in moist or wet soils in swampy wetlands, stream banks, and woodland habitats

Comments: *Erythranthe* is from the Greek *erythros* ("red") and *anthos* ("flower"), which refers to the color of the flowers. *Ringens* ("to gape") refers to the spreading mouth of the flowers. The flowers lack a floral scent and attract bumblebees, which can force their way past the narrow corolla throat, as pollinators. Winged monkey-flower (*Mimulus alatus*) is similar but has pinkish flowers borne on very short stems, squared stems with wings, and stalked leaves.

JACOB W. FRANK, NPS

WYOMING KITTENTAILS

Besseya wyomingensis
Plantain family (Plantaginaceae)

Description: Perennial, with hairy, upright stems 2"–12" tall. Alternate leaves are ¾"–2¾" long, covered with soft hairs, and elliptical to egg-shaped. The basal leaves have stems; the upper leaves are stalkless. Dense flower clusters are borne at the ends of the stems, with hairy 2-lipped (sometimes 3) flowers that lack petals and have 2 lavender to purplish stamens that project beyond the calyx. Fruit is a rounded capsule.

Bloom season: Mid-spring to midsummer

Range/habitat: Portions of western North America from British Columbia to Colorado in open grassy slopes

Comments: *Besseya* honors Charles Edwin Bessey (1845–1915), an American botanist and professor of botany at Iowa Agricultural College 1870–1884 and later on at the University of Nebraska. *Wyomingensis* ("of Wyoming") refers to the type locality of this species. Also known as Wyoming coral-drops.

EVAN M. RASKIN

GRAY BEARDTONGUE

Penstemon canescens
Plantain family (Plantaginaceae)

Description: Perennial, 1'–3' tall, clump forming. Leaves and stems are covered with short, white hairs. Lance-shaped to oblong leaves clasp the stem, are 2"–6" long, toothed along the margins, and opposite; basal leaves are stalked. Pale purple to pink (may also be white) trumpet-shaped flowers are 2-lipped, with 5 petals, and 1½" long; the sterile stamen is covered with hairs. Purple nectar guides line the corolla's throat. Fruit is a capsule.

Bloom season: Late spring to midsummer

Range/habitat: Eastern United States in rocky outcrops, dry slopes, and woodlands

Comments: *Penstemon* is from the Greek *pen* ("almost") and *stemon* ("thread") and refers to the stamens. *Canescens* ("ashy-gray hairs") refers to the fine hairs on the plant.

EATON'S PENSTEMON

Penstemon eatonii
Plantain family (Plantaginaceae)

Description: Perennial; 1 or several stout stems up to 3½' tall arise from a woody base. Lance-shaped leaves are mostly basal, dark green, thick, wavy along the margins, and up to 7" long; upper leaves are smaller. Upright spikes bear numerous red, tubular flowers, 1"–1½" long. Flowers may flare open at the tip and often hang somewhat downward. Fruit is a capsule.

Bloom season: Mid-spring through summer

Range/habitat: Southwestern United States and Rocky Mountain states in deserts, grasslands, rocky outcrops, woodlands, and shrublands

Comments: *Penstemon* is from the Greek *pen* ("almost") and *stemon* ("thread") and refers to the stamens. *Eatonii* honors David Cady Eaton (1834–1885), an American botanist and professor of botany at Yale, who wrote the descriptive text for *Beautiful Ferns*, published in 1886. Members of the Navajo tribe used this plant to treat tick bites and colic in livestock.

STEVE R. TURNER

LARGE BEARDTONGUE

Penstemon grandiflorus
Plantain family (Plantaginaceae)

Description: Perennial, 2'–4' tall, with round stems that may be white, green, or pale red. Basal leaves arise from a rosette and are egg-shaped, 2"–5" long, and bluish gray in color, with a somewhat fleshy texture. The fleshy stem leaves are egg-shaped, oppositely arranged, and also bluish gray to greenish gray. Large, tubular pinkish-lavender flowers are up to 2" long and borne in small clusters of 1–3 flowers from leaf axils. The 2-lipped flower, which appears partially flattened, has 2 upper lobes and 3 lower lobes. Within the flower's throat are purple veins that act as nectar guides. Fruit is a capsule.

Bloom season: Late spring to early summer

Range/habitat: Mostly central United States in meadows and open woodlands

Comments: *Penstemon* is from the Greek *pen* ("almost") and *stemon* ("thread") and refers to the sterile stamen. *Grandiflorus* ("large flowered") refers to the size of the flowers. Bumblebees and long-tongued bees are common pollinators of these flowers.

CRAIG MARTIN

PARRY'S PENSTEMON

Penstemon parryi
Plantain family (Plantaginaceae)

Description: Perennial, 2'–4' tall. From a basal rosette of arrow-shaped leaves arise smooth stems that bear narrow, pale green lance-shaped leaves, 2"–5" long, which are opposite in their arrangement. The lower portion of the stem and some of the leaves may have a purplish cast. Long spikes of tubular, pink flowers, ¾"–1" long, are borne in the upper portion of the plant. The 2-lipped flowers have 5 lobes; each lobe bears a pink to purplish stripe down the middle. The mouth of the flower has fine white hairs. Fruit is a capsule.

Bloom season: Early to mid-spring

Range/habitat: Southern Arizona and northern Mexico in desert canyons

Comments: *Penstemon* is from the Greek *pen* ("almost") and *stemon* ("thread") and refers to the sterile stamen. *Parryi* honors Charles C. Parry (1823–1890), who served on the US–Mexican Boundary Survey (1848–1855) as a surgeon and botanist. Later on, he collected many plants new to science in Utah, Colorado, and other western states. The hot pink flowers attract hummingbirds and long-tongued insects as pollinators. This and other penstemons were moved from the Figwort family (Scrophulariaceae) to the Plantain family.

CLIFF BEARDTONGUE

Penstemon rupicola
Plantain family (Plantaginaceae)

Description: Perennial, up to 4" high; mat forming. Smooth gray-green leaves are round or oval, ¼"–¾" long, oppositely arranged, finely toothed along the margins, and covered with a fine waxy coating. The light purple to deep pink flowers are about 1½" long and tubular. The 5-lobed flowers flare open at the mouth. Fruit is a capsule.

Bloom season: Late spring through summer

Range/habitat: Pacific Northwest in rocky habitats such as cliffs, talus slopes, and rocky areas

Comments: *Penstemon* is from the Greek *pen* ("almost") and *stemon* ("thread") and refers to the sterile stamen. *Rupicola* ("growing on ledges or cliffs") refers to where this plant is often found growing. Also called rock penstemon.

SEA THRIFT

Armeria maritima

Plumbago family (Plumbaginaceae)

Description: Perennial, 6"–12" tall; in mound-forming clusters. Linear basal leaves are stiff and spiny and form dense, mat-like clusters about 4" tall. Leafless flowering stalks bear rounded clusters, ½"–1½" wide, of tiny, pink or lavender (sometimes white) flowers. Papery purplish bracts subtend the flower clusters; petals are also papery and may last on the plant long after the blooming season. Fruit is a single seed enclosed within a bladder that is often enclosed by the sepals.

Bloom season: Mid-spring to summer

Range/habitat: Circumboreal distribution includes western United States, northern North America, and Greenland on coastal bluffs, headlands, and beaches.

Comments: *Armeria* is from the French name *armoires* ("cluster-headed *Dianthus*"), which refers to a different, although similar-looking plant. *Maritima* ("coastal") refers to the plant's distribution. Also known as sea pink or thrift, this hardy plant is often used in coastal or inland gardens.

SKYROCKET GILIA

Ipomopsis aggregata

Phlox family (Polemoniaceae)

Description: Biennial or short-lived perennial. One to several flowering stalks arise 1'–3' from a basal rosette of highly dissected leaves. Stem leaves are smaller. Flowers are borne in loose clusters and are mostly red, although orange or yellow forms exist. Flowers are ¾"–1½" long, with white to yellowish sepals; the corolla tube flares open to form a 5-pointed star. Fruit is a capsule.

Bloom season: Summer

Range/habitat: Western North America in dry meadows, roadsides, rocky outcrops, or lightly wooded areas

Comments: *Ipomopsis* is from the Greek *ipo* ("to strike") and *opsis* ("resembling"), in reference to the striking flowers. *Gilia* honors Filippo Luigi Gilii (1756–1821), an Italian astronomer and coauthor of *Observationi Fitologicha* (1789–1792). *Aggregata* ("aggregated") refers to the cluster of basal leaves; the leaves may have a bit of a foul, skunk-like odor. Hummingbirds, butterflies, and certain moths are the primary pollinators that can reach the flower's nectar. Also known as scarlet gilia, the plant is highly variable throughout its range.

ALEX HEYMAN

MUSTANG CLOVER

Leptosiphon montanus
Phlox family (Polemoniaceae)

Description: Annual, with thin stems up to 2' tall. Hairy compound leaves are divided into needlelike linear lobes, ¾"–1" long. Showy flowers have a red, 1"-long, tubular corolla, which flares at the opening to form 5 white to pink flat lobes. Petals sometimes white. The base of the lobes is white or yellow, with a reddish to purplish spot or band. Fruit is a seed.

Bloom season: Mid-spring to midsummer

Range/habitat: Endemic to California in oak woodlands at higher elevations

Comments: *Leptosiphon* is Greek for "slender tube," a reference to the corolla shape. *Montanus* ("of the mountains") refers to the plant's distribution. Often blooming in profusion, this plant has been used in horticultural settings as a ground cover.

STEVE R. TURNER

SMOOTH PHLOX

Phlox glaberrima
Phlox family (Polemoniaceae)

Description: Perennial, 2'–4' tall, clump forming. Thin, lance-shaped leaves are up to 4" long, opposite, and lack hairs. Flowers are borne in pyramid- to dome-shaped clusters containing 3–20 flowers. The 5-lobed flowers are rose to reddish purple, about 1" wide, and with a long corolla tube. Fruit is a narrow seed capsule.

Bloom season: Spring

Range/habitat: Southeastern and central United States in stream banks, swamps, marshes, prairies, moist meadows, and woods

Comments: *Phlox* ("flame") is in reference to the flower color of certain members of this genus. *Glaberrima* ("without hairs") describes the stems and leaves, which are hairless. The fragrant flowers attract moths and butterflies as pollinators.

ELEANOR DIETRICH

PROCESSION FLOWER

Polygala incarnata
Milkwort family (Polygalaceae)

Description: Annual, 12"–24" tall, with smooth stems. Alternate leaves are linear, up to ½" long, and are upright or appressed to the stem. The pinkish flower is about ½" long. Petals are lobed at the tip, and the 3 petals are united to form a slender tube. Fruit is an egg-shaped capsule with small seeds that have an attached air-filled sac.

Bloom season: Mid-spring to early fall; mainly summer

Range/habitat: Eastern North America in prairies, grasslands, meadows, and open fields.

Comments: *Polygala* is from a Greek word meaning "much milk" and refers to the ancient idea that cattle that fed on these plants produced higher quantities of milk. *Incarnata* ("flesh colored") refers to the pinkish flowers. Because the leaves are small and appressed to the stem, the plant appears to have no leaves at all, which leads to another common name: slender milkwort. The common name procession flower comes from the flowers woven into garlands to celebrate the fifth Sunday after Easter.

ALEX HEYMAN

WATER SMARTWEED

Persicaria amphibia
Buckwheat family (Polygonaceae)

Description: Perennial, growing in aquatic habitats or terrestrial areas that periodically become inundated. Rhizome produces a thick stem that, depending on the habitat, may float, creep along the ground, or grow upright. Stems, which may be up to 10' long in aquatic forms, send out roots from nodes. Lance-shaped leaves are borne on long stems and may be 2"–8" long and ½"–3" wide. A leafless flowering stalk bears an elongated cylindrical cluster of tiny, 5-lobed, pinkish flowers. Each flower is about ¼" long and bears 5 pinkish stamens that project beyond the flower's petals and a divided style. Flowers may be perfect or unisexual (male or female). Fruit is a seed.

Bloom season: Midsummer to fall

Range/habitat: Native to much of North America in ponds, wetlands, streams, and marshes

Comments: *Persicaria* is from the Latin *persica* ("peach") and *aria* ("pertaining to"), in reference to the leaves of smartweeds resembling those of peaches. *Amphibia* ("amphibian") refers to the plant's rooting habits of either in water or on land. Also known as longroot smartweed or amphibious bistort for its growth in both aquatic and terrestrial habitats.

ANNIKA LINDQUIST

SUNBRIGHT

Phemeranthus parviflorus
Purslane family (Portulaceae)

Description: Perennial, about 4" tall. Succulent leaves are linear and about 1"–2" long. A leafless flowering stalk bears a small cluster of reddish-pink or reddish-purple star-shaped flowers that are ⅓"–½" wide. Flowers have 5 pink petals and 5–10 pink stamens with yellow anthers. Fruit is an upright oval capsule, ⅛" long.

Bloom season: Late spring through summer

Range/habitat: Central United States and northern Mexico in thin soils in grasslands, canyon washes, rocky bluffs, and dry woodlands

Comments: *Phemeranthus* has an unclear origin. The Greek *ephemer* ("for a day") and *anthus* ("flower") focus on the short-lived flowers, but *phem* ("to report") could point to *Pheme*, the Greek goddess of fame. *Parviflorus* ("with small flowers") refers to the clusters of small flowers. Each flower opens for only a couple of hours during one afternoon before dropping its sepals and petals. Also known as prairie fame-flower or rockpinks due to the flower color and the plant's habit of growing on rocky surfaces.

STEVE R. TURNER

DARK-THROATED SHOOTING STAR

Dodecatheon pulchellum
Primrose family (Primulaceae)

Description: Perennial; single leafless flowering stems grow upright to 16" tall. Basal leaves are inversely lance-shaped to oval, ¾"–2" wide and 1"–10" long, and smooth or slightly toothed along the margins. The starlike flowers have a purple-spotted calyx that has 5 short lobes. Flower clusters may contain a few or many flowers. Corolla has 5 deep pink or purplish petals that curve backward from a short, yellow tube. The tube often has a wavy purplish line around its base, and the base of each petal has a white spot. The stamen's yellow filaments are joined together, forming a short tube, and topped by yellow or purplish anthers. Fruit is a capsule.

Bloom season: Mid-spring to late summer

Range/habitat: Western North America in moist coastal sites (may be saline), seeps, wetlands, stream banks, and mountain meadows

Comments: *Dodecatheon* is from the Greek *dodeka* ("twelve") and *theos* ("god"), referring to the twelve Olympian gods, who were thought to protect primroses. *Pulchellum* ("pretty" or "beautiful") describes the flowers. Bumblebees pollinate the flowers by hanging upside down from the inner ring and "buzzing" their wings to shake loose the pollen. Also known as few-flowered shooting star, for the number of flowers and their appearance.

PATRICK ALEXANDER

PARRY'S PRIMROSE

Primula parryi

Primrose family (Primulaceae)

Description: Perennial, 9"–20" tall, with a skunk-like odor. Basal rosettes of lance- to egg-shaped leaves often are clumped together; leaves are up to 12" long. A cluster of 5–25 purplish flowers are arranged along an elongated stalk. The 5 petals are fused below into a short tube before flaring open, revealing a yellow central ring. Flowers are about 1" wide. Fruit is a capsule.

Bloom season: Late spring to late summer

Range/habitat: Rocky Mountain states along streams, springs, and moist mountain meadows at high elevations

Comments: *Primula* is from the Latin *primus* ("first") and *ulus* ("tiny"), in reference to the flowers opening in early spring as soon as the snow melts. *Parryi* honors Charles C. Parry (1823–1890), an English physician and botanist who collected plants widely in the midwestern and western United States, serving as a physician/botanist on the US–Mexican Boundary Survey. Parry was the first US Department of Agriculture botanist (1869–1871) and, due to his extensive collecting in Colorado, was nicknamed King of Colorado Botany. The carrion-like aroma attracts flies as pollinators.

CAVE PRIMROSE

Primula specuicola

Primrose family (Primulaceae)

Description: Perennial, 2"–11" tall, with withered leaves at the base. Leaves are ¾"–8" long, spatula-shaped to elliptical, and variously toothed along the margins. Leaves are green above and white-mealy below. Lavender to pink flowers with the corolla tube rimmed in yellow are ½"–⅔" wide and arranged in clusters of 5–40 flowers at the end of a leafless stalk. Fruit is a capsule.

Bloom season: Late winter to early summer

Range/habitat: Seeps and hanging gardens in the Four Corners region of the US Southwest

Comments: *Primula* is from the Latin *primus* ("first") and refers to the early blooming period of the flowers. *Specuicola* means "cave inhabiting" and refers to the habitat where these plants occur. The fragile seep habitat where these plants grow might consist of some minimal soil or fractures in the rock surface that the roots penetrate. Remember to view the plants from a distance to not disturb this sensitive habitat. Also known as Easter flower, because the plants flower around Easter time.

CLAIRE WEISER

RED COLUMBINE

Aquilegia formosa
Buttercup family (Ranunculaceae)

Description: Perennial, up to 3' tall. Leaves are twice divided into 3s; the majority of the leaves are basal. The red and yellow, drooping flowers have 5 long spurs and a central cluster of yellowish stamens and styles that project beyond the flower's mouth. Sepals are the reddish or orange spreading outer parts; the petals are the yellow inner parts. Fruit is a 5-sectioned capsule that splits open at maturity to release tiny black seeds.

Bloom season: Late spring to late summer

Range/Habitat: Western North America in moist woodlands, along stream banks, and near seeps

Comments: *Aquilegia* ("eagle") refers to the talon-like spurs on the flowers. *Formosa* ("beautiful") refers to the flower color and shape. These flowers attract hummingbirds, sphinx moths, and butterflies as pollinators, which use their long tongues or proboscises, respectively, to reach the nectar at the bulbous base of the spurs. Some Native Americans pinched off the spurs and sucked out the nectar as a sweet treat. In contrast, the seeds are toxic. Serpentine columbine (*A. eximia*) is endemic to California and has larger flowers.

CHUCK TAGUE

LEATHER-FLOWER

Clematis viorna
Buttercup family (Ranunculaceae)

Description: Perennial vine; stems up to 12' long, woody at the base. Compound leaves have 4–8 oval to lance-shaped leaflets, which may again be divided or lobed. The terminal leaflet is a tendril. Overall size of the leaves varies from ¾" to 4¾". The lavender to reddish-brown leathery flowers are bell- or urn-shaped, lack petals, and may be cream at the tips. Sepals are recurved at the tip. Styles are hairy. Fruit is a seed with a long, feathery tail.

Bloom season: Spring to fall

Range/habitat: Southeastern United States in thickets, woods, and stream banks

Comments: *Clematis* is the Greek name of a climbing plant. *Viorna* is derived from the Latin *viburno* ("viburnum"). The plants are toxic. The flowers are tough and leathery; hence the common name. Also known as vasevine.

STEVE R. TURNER

ROUND-LEAF HEPATICA

Hepatica americana
Buttercup family (Ranunculaceae)

Description: Perennial, up to 9" tall. Leaves are borne on lightly hairy stems, are up to 6" long with 3 rounded lobes, and remain attached until the following spring. Flowers range from lavender to pink or white, arise on 4"–6" hairless stems, and are ½"–1" wide. There are 5–12 rounded petals (generally 6) forming a greenish center with numerous white stamens and 3 hairy sepals. Fruit is a seed with hairs.

Bloom season: Spring to midsummer

Range/habitat: Eastern North America woodlands, bluffs, and rocky outcrops in acidic soils

Comments: *Hepatica* refers to the 3-lobed leaf resembling the shape of the liver. *Americana* ("of America") refers to the plant's distribution. The leaves, which don't appear until the flowers start to bloom, turn dark green or brown in summer and remain attached until the following spring. Perhaps due to the liver-like shape of the leaves, settlers made an herbal infusion from the leaves to treat liver ailments. Sharp-lobed hepatica (*H. acutiloba*) is similar but has sharp points on the leaf lobes and grows in wetter sites that are lower in acidity. Some authors place these species in the *Anemone* genus.

STEVE R. TURNER

QUEEN OF THE PRAIRIE

Filipendula rubra
Rose family (Rosaceae)

Description: Perennial, 6'–8' tall, with (sometimes) red stems. Leaves are palmately or pinnately compound and 2'–3' long. There are generally 7–9 lance-shaped coarsely toothed leaflets, each about 6" long, and 1 large terminal leaflet that has 7–9 lobes and is 4"–8" long. Flowering clusters are composed of numerous (200–1,000) tiny, fragrant pinkish flowers; each flower is about ⅓" wide. The clusters are 5"–8" wide. Exserted stamens give the flower clusters a "fuzzy" appearance. Fruit is a seed.

Bloom season: Summer

Range/habitat: Northeastern and central United States and southern Canada in swamps, meadows, wetlands, fens, seeps, and wet prairies

Comments: *Filipendula* is from the Latin *filum* ("a thread") and *pendulus* ("hanging"), for the root tubers hanging together by fine rootlets. *Rubra* ("red") refers to the flower color. The fragrant spray of flowers may be pollinated by sweat bees or other small insects, or by the wind carrying pollen grains to nearby plants. Native peoples used the roots to treat a variety of ailments, including diarrhea, dysentery, and heart problems; it was also used as an aphrodisiac. Also called meadowsweet.

PRAIRIE-SMOKE

Geum triflorum

Rose family (Rosaceae)

Description: Perennial, 4"–20" tall. Basal leaves are fernlike and hairy. The 1'-long, vase-shaped flowers are arranged in groups of 3 and hang downward. The pink to reddish (sometimes yellowish) petals are hidden beneath 5 red sepals and curved bracts. Fruit is a seed with a long, hairy plume that resembles a feather duster.

Bloom season: Late spring to midsummer

Range/habitat: Western and northern North America in rocky outcrops, grassy slopes, moist meadows, and prairies

Comments: *Geum* is the classical Latin name for this genus. *Triflorum* ("three flowers") refers to the flower clusters. The plant was first collected for science by Meriwether Lewis in 1806. The 2"-long, feathery tails on the seeds, which resemble whiskers, help the seeds be dispersed by the wind and give the plant its other common names: old man's whiskers and lion's mane, for the long hairs on the seeds. Native peoples boiled the roots to make a tea for sore throats.

JIM FOWLER

SCARLET CINQUEFOIL

Potentilla thurberi

Rose family (Rosaceae)

Description: Perennial, 1'–2½' tall, clump forming. Plants have palmately compound leaves borne on stems that may reach 4" long. Each of the 5–7 leaflets is 1"–2" long, egg- to lance-shaped, and toothed along the margins. The leafy flowering stems may be 30" tall and bear cup-shaped, rose-red flowers with 5 petals that are rounded at the tip, 5 greenish sepals with narrow bracts (called epicalyx bracts) between the sepals, and reddish stamens. Petals are darker toward the center. Fruit is a capsule.

Bloom season: Summer

Range/habitat: Southwestern United States (primarily Arizona and New Mexico) and northern Mexico along stream banks and roadsides and in moist meadows and coniferous woodlands

Comments: *Potentilla* is from the Latin *poten* ("powerful"), referring to the medicinal value of the plant. *Thurberi* honors George Thurber (1821–1890), a botanist who served on the US–Mexican Boundary Survey and collected numerous plants new to science in the southwestern United States, such as organ pipe cactus (*Stenocereus thurberi*). Also known as red cinquefoil or Thurber's cinquefoil; "cinquefoil" refers to the 5-leaved foliage.

JIM FOWLER

NORTHERN PITCHER PLANT

Sarracenia purpurea
Pitcher-Plant family (Sarraceniaceae)

Description: Perennial carnivorous plant with hollow and inflated leaves known as "pitchers." Pitcher-like leaves are up to 12" tall, with an erect triangular lid (called a hood), which may have intricate purplish veins (some leaves are just green, especially when young). The hood shades the opening and has 2 large crescent-like wings on either side of the lip. There are also stiff hairs along the inside of the hood, pointing downward. These pitchers may last for two seasons. Red to purplish flowers (greenish in parts of the range) arise on leafless stalks, 1'–2' above the basal leaves, and arch downward when mature. The maroon petals are rounded; the reddish sepals are pointed, and the large sepal is umbrella-shaped. Fruit is a knobby seed.

Bloom season: Mid-spring to early summer

Range/habitat: Across much of Canada, the northeastern United States, portions of the southern United States, and parts of the Pacific Northwest in sphagnum bogs, fens, and ditches that hold water for long periods of time

Comments: *Sarracenia* honors Michel Sarrazin (1659–1734), a French physician who immigrated to New France and became a physician and plant collector in Quebec. *Purpurea* ("purple") refers to the flower color. This *Sarracenia* obtains nutrients from prey that becomes trapped within the large pitcher and is then dissolved by digestive enzymes. The plant produces these enzymes in the first year, but during the second season, communal microorganisms feed on the remains of decaying insects, and some of these nutrients may be consumed by the plant. In addition, larval forms of a mosquito and midge coexist within the pitcher's fluid, feeding on the microorganisms and unaffected by the digestive juices. This plant is the floral symbol of the eastern Canadian province of Newfoundland and Labrador.

STEVE R. TURNER

DAKOTA VERVAIN

Glandularia bipinnatifida
Verbena family (Verbenaceae)

Description: Annual or short-lived perennial; highly variable in terms of height or spreading along the ground. The 4-angled stems are generally 6"–18" long but may reach 2'. Stems and leaves are covered with white hairs. Opposite leaves are up to 2½" long, twice (sometimes 3 times) pinnate, and each segment is finely divided into linear lobes. The pink to purplish flowers are borne in rounded clusters, and each flower has 5 lobes surrounding a white center; each lobe is notched at the tip. Awl-shaped bracts subtend the flowers. Fruit is 4 nutlets.

Bloom season: Early spring to fall

Range/habitat: Widespread across southern half of the United States from California in fields, grasslands, prairies, and shrublands

Comments: *Glandularia* ("glandular") refers to the glands located on the calyx. *Bipinnatifida* ("twice pinnately divided") describes the leaves. Native Americans used the leaves to treat snakebites. Also known as prairie vervain, Dakota mock vervain, and *Moradilla*, a Spanish name that translates to "little purple one." Southwest mock vervain (*G. gooddingii*) is similar but with lobed leaves and grows in the southwestern United States and northern Mexico.

JIM FOWLER

ARROWLEAF VIOLET

Viola sagittata
Violet family (Violaceae)

Description: Perennial; stems up to 6" long. Basal leaf blades are 1½"–4" long, narrowly triangular to spearhead-shaped, slightly toothed along the margins, and occasionally with fine hairs along the margins. Leaves are longer than wide and have small lobes at the base. Hairless flowering stalks arise from the base of the plant, and the nodding flower is ¾"–1" wide, with 5 purple-violet petals. The lower and lateral petals have fine white hairs at their base; the lower central petal has a white spot with purple veins. A short, slightly curved nectar spur forms behind this lower petal. Fruit is a hanging, ovoid-shaped capsule.

Bloom season: Mid-spring to midsummer

Range/habitat: Eastern North America in moist prairies, meadows, fields, riverbanks, and woodlands

Comments: *Viola* is the Latin name for various sweet-smelling flowers, including violets. *Sagittata* is derived from the Latin *sagitta* ("an arrow"), which refers to the leaf shape. Later in summer, the plant may produce self-fertilizing (cleistogamous) flowers that lack petals and may produce viable seeds. The plants are an early successional species after a fire as dormant seeds sprout to colonize the disturbed ground.

WHITE FLOWERS

This section includes flowers that range from creamy to bright white. Some plants produce flowers that tend toward pale yellow or pale green, so if you don't find your flower here, check those sections.

STEVE R. TURNER

HAIRY WILD PETUNIA

Ruellia humilis
Acanthus family (Acanthaceae)

Description: Perennial, 1½'–2' tall; stems are covered with fine white hairs. Oblong to linear-shaped leaves are 4" long, opposite, hairy, with smooth margins. Leaf stems are also hairy. Funnel-shaped blue to lavender flowers are 1¼"–3" long; the 5 petals end with shallow, rounded lobes that are ruffled. Short calyx has 5 linear teeth and is hairy. Fruit is a capsule.

Bloom season: Late spring to mid-fall

Range/habitat: Eastern and central United States in prairies, fields, glades, and open woods

Comments: *Ruellia* honors Jean de la Ruelle (1474–1537), a French physician and herbalist who was the King of France's physician from 1515 to 1547. *Humilis* ("low growing" or "dwarf") refers to the height of this plant, which may appear bushy because of the multibranch stems. Flowers last for 1 day, opening in the morning and falling off in the evening. Long-tongued bees are common pollinators of these flowers.

JUDY PERKINS

OUR LORD'S CANDLE

Hesperoyucca whipplei
Century-Plant family (Agavaceae)

Description: Perennial. Sharp-pointed leaves, 8"–40" long, are finely toothed along the edges. Thick flowering stalk arises from the basal cluster of leaves, reaching 3'–10' (or more) in height. Flower cluster may number in the hundreds of white or purplish, 1¼"-wide, bell-shaped flowers. Fruit is a capsule.

Bloom season: Late spring to early summer

Range/habitat: California and extreme southwestern Arizona into Baja California in coast shrublands, chaparral, Joshua tree forests, and oak or pinyon woodlands

Comments: *Hesperoyucca* is from the Greek *hespero* ("western") and *yucca* ("*Yucca*-like plant"), which describes the plant's range and appearance. *Whipplei* honors Amiel Wicks Whipple (1818–1863), a topographical engineer who served on the US–Mexican Boundary Survey in 1853–1856 and who was the leader of the Pacific Railroad Survey Expedition along the 35th parallel. The plants take several years (often more than 5) to reach maturity; the plant dies after flowering. The flowers are pollinated by a female yucca moth species as she transfers pollen to the stigma then lays her eggs into the stigma's ovary. These fire-adapted plants may also resprout after a fire event. Native peoples used the leaves for cordage; the young flowering stalks were cut and harvested for food. Also known as chaparral yucca or Quixote yucca.

STEVE R. TURNER

WAPATO

Sagittaria latifolia
Water-Plantain family (Alismataceae)

Description: Perennial, typically 1'–4' long, but may have extensive branching. Large arrowhead-shaped leaves are up to 10" long; while submerged, leaves are either lance-shaped or linear. The flowering stalks bear white flowers that are ½"–1" wide and have 3 greenish sepals and 3 white petals. Male flowers are ball-shaped, smaller, and located below the female ones. Fruit is a sharp-beaked, winged seed borne in a rounded cluster.

Bloom season: Mid- to late summer

Range/habitat: Across most of North America in ponds, marshes, stream edges, and lakes

Comments: *Sagittaria* is from the Latin word *sagitta* ("an arrow") and *latifolia* ("broad-leaved"), referring to the large arrowhead-shaped leaves. Native Americans still harvest the thick tubers, which are eaten baked. Native women often collected the potato-like roots by suspending themselves from a canoe, digging the roots out of the soft mud with their toes, then collecting the roots as they floated to the surface. Waterfowl and muskrats eat the roots and tubers. The seeds eventually sink in the pond and germinate underwater. "Wapato" is the Chinook word for this plant, also known as common arrowhead.

MARGARET MARTIN

TEXTILE ONION

Allium textile
Amaryllis family (Amaryllidaceae)

Description: Perennial; plants form from a ¾"-thick bulb. Threadlike leaves, 2–4 per flowering stalk, are round; can be up to 9" long. Leafless flowering stalks, 1"–9" long, bear rounded clusters of 5–55 white or pale pink flowers. Bell- or urn-shaped flowers have 6 tepals, each with deep purple nectar guides. Fruit is a capsule.

Bloom season: Mid-spring to midsummer

Range/habitat: Central North America from Saskatchewan to Arizona and New Mexico in sandy desert openings, shrublands, prairies, and woodlands

Comments: *Allium* is the Greek name for garlic. *Textile* ("textile" or "fabric") refers to the brown fabric-like fibers that loosely encase the bulb. The very similar looking large-petaled onion (*A. macropetalum*) has 3–5 veins on the flower bracts, while textile onion has 1. When conditions are right, these plants may carpet large expanses.

JIM FOWLER

RAMP

Allium tricoccum
Amaryllis family (Amaryllidaceae)

Description: Perennial from a bulb; flowering stems up to 18" tall. Lance-shaped leaves are 7½"–12" long, somewhat wide, and wither before the flowering stalk appears. A hemispherical cluster of white to creamy white flowers forms at the tip of a 6"- to 18"-long flowering stalk. The head is 1"–2" wide and composed of 20–40 flowers that have 6 tepals and 6 exserted stamens. Fruit is a 3-celled capsule; each cell contains 1 seed.

Bloom season: Summer

Range/habitat: Eastern North America in floodplains, woodlands, and wooded bluffs

Comments: *Allium* is the Greek name for garlic. *Tricoccum* ("three seeds") refers to the number of seeds produced inside each capsule. Throughout Appalachia, the leaves and bulbs are harvested in spring for consumption during annual Ramp Festivals or for farmers' markets; the scallion-like flowering stalks are also edible. The Illinois Indians called these plants "*Chicagoua,*" from which the name Chicago was derived. The common name ramp is because these plants resembled an English plant called ransom (*A. ursinus*) to early settlers. Also known as wild leek or wood leek.

ELEANOR DIETRICH

SWAMPY LILY

Crinum americanum
Amaryllis family (Amaryllidaceae)

Description: Perennial; from a bulb with clump-forming stems. Long strap-like leaves are 2'–4' long and 2"–3" wide. Flowering stems are 2'–3' tall and bear a cluster of 2–6 flowers at the top. The 6 white, symmetrical sepals and petals may be tinted with pink and are 3"–5½" long, curving backward to give the flower a ball-like shape and expose the pink to reddish stamens. Fruit is a capsule.

Bloom season: Summer; may bloom throughout the year in various locations

Range/habitat: Southeastern United States in swamps, wetlands, marshes, and stream banks

Comments: *Crinum* is from the Greek *krinon* ("lily"). *Americanum* ("of America") defines the distribution of this plant. Also known as string lily, spider lily, or seven sisters lily, for the flower cluster resembling the seven daughters of Atlas and Pleione in the Pleiades star cluster.

ALY M. HARMON

WOODLAND SPIDER LILY

Hymenocallis occidentalis

Amaryllis family (Amaryllidaceae)

Description: Perennial, 1½'–2' tall; from a bulb. Lance-shaped leaves are up to 2' long and ¾" wide, fleshy, and often wither prior to flowering. Sweet-smelling funnel-shaped flowers, 3–9, are borne atop a 1'- to 1½'-long flowering stalk. The green perianth tube is ¾"–1½" long, and the flowers are about 6" across. Individual flowers are spiderlike, with a cuplike center (the body) and 6 strap-like appendages (the legs). Fruit is an oval to spherical capsule.

Bloom season: Summer

Range/habitat: Southern United States in open woodlands or mature forests

Comments: *Hymenocallis* is from the Greek *hymen* ("membrane"). *Kallos* ("beauty") refers to the membrane that connects the stamens together. *Occidentalis* ("western") refers to the plant's distribution in Texas. Also called hammock spider lily or northern spider lily.

STEVE R. TURNER

FALSE GARLIC

Nothoscordum bivalve

Amaryllis family (Amaryllidaceae)

Description: Perennial; the 1–2 flowering stems are 8"–16" tall, from a bulb. Narrow, basal leaves (1–4) are 4"–15" long, linear, and partially truncated at the tip. Whitish flowers are borne in small clusters (generally 4–8) that are flat-headed and have 2 bracts subtending each cluster. Individual flowers have 6 tepals with a yellow base and reddish-brown or green midvein stripe on the undersides. About ½" wide with bright yellow stamens, flowers do not produce an aroma. Fruit is a capsule.

Bloom season: May bloom in early to late spring and again in fall

Range/habitat: Southern United States from Arizona to Virginia and south into South America in lawns, meadows, prairies, savannas, and thin woods

Comments: *Nothoscordum* is from the Greek *nothos* ("false") and *skordon* ("garlic"), which refers to the false garlic nature of this plant. *Bivalve* ("two valves") refers to the 2 bracts that subtend the flower cluster. The lack of an onion or garlicky odor to the flowers or leaves is reflected in the common name. Bees and smaller butterflies are the primary pollinators of these flowers. Also known as crow poison; however, the plant doesn't have any toxic properties.

TYLER CANNON

EVENING-STAR RAIN-LILY

Zephyranthes drummondii
Amaryllis family (Amaryllidaceae)

Description: Perennial, up to 12" tall, from a round bulb. Gray-green, grasslike leaves are up to 12" long or longer, ¼" wide, and are more common when the plant is not in flower. A solitary, starlike flower with 6 petals is borne on a leafless flowering stalk that starts out pinkish but turns green at maturity. Fruit is a 3-lobed capsule.

Bloom season: Summer

Range/habitat: South-central United States in prairies, savannas, and woodlands

Comments: *Zephyranthes* is from Zephyrus, Greek god of the west wind. *Drummondii* is in honor of Thomas Drummond (1790–1835), a Scottish botanist who collected plants extensively in Texas but unfortunately died in 1835 while on a plant-collecting exploration in Cuba. Also known as evening rain-lily, or *cebolleta*, which means "onion" in Spanish and refers to the onion-like bulbs. Flowering depends on summer rainfall; hence the common name. Some taxonomists consider this to be in the *Cooperia* genus. Also known as Drummond's rain lily.

STEVE R. TURNER

HAIRY ANGELICA

Angelica venenosa
Carrot family (Apiaceae)

Description: Perennial, 3'–8' tall. Smooth stems are light purple or have purple spotting. Alternately arranged leaves are compound and borne on short stems. Basal blades are up to 24" long and triangular to egg-shaped in outline. Blades are pinnately divided (sometimes a second time); oval to egg-shaped leaflets are ⅓"–1" long, blunt or pointed at the tip, and finely toothed along the margins. Undersides have short, dense hairs. Upper leaves are smaller and more rounded in outline. Flowers are arranged in semi-hemispherical umbels in clusters that range from 8 to numerous flowers; individual flowers are white and tiny, with 5 white petals and 5 stamens. Flowering stalks and stems are hairy. Fruit is a flattened seed.

Bloom season: Late spring to midsummer

Range/habitat: Eastern North America in dry sandy or rocky fields, prairies, savannas, stream banks, and woodlands

Comments: *Angelica* ("angel") refers to the medicinal and angelic protection against evil properties as revealed to humans by an archangel. *Venenosa* ("poisonous") refers to the toxic roots. The flowers attract a wide range of pollinators, from small flies to beetles and bees. The common name is in reference to the short hairs on the undersides of the leaves and toward the ends of the stems.

STEVE R. TURNER

WATER HEMLOCK

Cicuta maculata
Carrot family (Apiaceae)

Description: Short-lived perennial or biennial with hollow stems, 1½'–6' tall. The long leaves are doubly compound and up to 16" long. Lance-shaped leaflets are 1"–4" long and have a wedge-shaped base and pointed tip; leaflets are rarely lobed, and the purplish veins extend to the base of the notches between the teeth along the margins, not the tips. Tiny white flowers are borne in dense clusters made up of 10–20 smaller, dome-shaped umbels of flowers. Overall clusters may be up to 6" wide. Individual flowers are ⅛" wide and have 5 white petals. Fruit is a rounded, corky capsule.

Bloom season: Midsummer to fall

Range/habitat: Most of North America from Canada to Mexico in wet soils in disturbed ground, along ditches and stream edges, ponds, and marshes

Comments: *Cicuta* is a Latin name used by Pliny for this or another similar-looking toxic plant. *Maculata* ("spotted") refers to the purplish spots on the stems. The white tuberous roots concentrate cicutoxin, which is poisonous. Water hemlock is considered by many to be the most toxic plant in North America. Also known as cowbane, for the toxic effect on cattle that consume the fernlike foliage. Poison hemlock (*Conium maculatum*) is the plant used to kill the Greek philosopher Socrates.

ELEANOR DIETRICH

RATTLESNAKE MASTER

Eryngium yuccifolium
Carrot family (Apiaceae)

Description: Perennial, 4'–5' tall. Basal leaves are sword-shaped, up to 3' long, with parallel venation, and edged with bristles. Greenish-white flowers are tiny but arranged in spherical clusters about 1" wide that arise in umbels. Flowers resemble thistle flower heads. Whitish, pointed bracts subtend the flower heads. Fruit is a seed with serrated wings to aid in dispersal.

Bloom season: Late spring through summer

Range/habitat: Central and eastern United States in rocky outcrops, prairies, savannas, and edges of swamps and seeps

Comments: *Eryngium* may refer to the prickly or spiny nature of the plant. *Yuccifolium* ("*Yucca*-like leaves") refers to the stiff, large leaves, which resemble those of *Yucca* species. This plant was used to treat rattlesnake bites; hence its common name. When mature, the flowers develop a bluish tint. A variety of insects pollinate the flowers. Caterpillars of the rattlesnake-master stem-borer moth (*Papaipema eryngii*) burrow into the roots and stems to pupate. Also known as button snake-root.

COW PARSNIP

Heracleum maximum
Carrot family (Apiaceae)

Description: Perennial, 3'–9' tall, with hollow stems, often in clusters. Large compound leaves have 3-lobed segments with toothed margins; each segment is 6"–16" long and nearly as wide. New leaves have white-woolly hairs. Numerous small white flowers are borne in terminal umbrellalike clusters. Smaller side clusters contain fewer flowers. Fruits are large, flat seeds.

Bloom season: Late spring through summer

Range/habitat: Across much of North America except the most southern states (from Texas to Florida) in moist sites along stream banks, avalanche paths, roadsides, or clearings

Comments: *Heracleum* is in honor of Heracles (Hercules) and refers to the largeness of the leaves, flowers, and overall plant. *Maximum* ("largest") refers to the size of the leaves. Sometimes called Indian celery or Indian rhubarb because coastal tribes ate the peeled young stems like celery. Deer, elk, bear, and mountain sheep also forage on the stems and leaves. The flowers attract a wide range of insects as pollinators.

HEMP DOGBANE

Apocynum cannabinum
Dogbane family (Apocynaceae)

Description: Perennial, 1'–3' tall, with stout, upright stems. A milky sap exudes when the stem or leaves are cut. Upright leaves are elliptical to egg-shaped, 2"–3" long, arranged oppositely along the stem, and have pointed tips. Small white or green bell-shaped flowers, ⅛"–¼" wide, arise on short stalks in clusters along the stem. Fruit is a long, thin pod.

Bloom season: Summer

Range/habitat: Widespread throughout North America in meadows, disturbed areas, forest openings, coastal areas, and shrublands

Comments: *Apocynum* is from the Greek words *apo* ("away") and *kyon* ("dog"), although the derivation is confusing; the plant may have been used to poison wild dogs. *Cannabinum* is from the Greek word meaning "hemp"; hence the common name, as well. Native Americans used the strong fibers of this plant to make cordage for fishing nets, snares, and sewing and as string. The milky latex may cause skin blistering, and although other parts of the plant are toxic, these are host plants for the larvae of snowberry clearwings (*Hemaris diffinis*) and hummingbird clearwings (*Hemaris thysbe*), small moths that resemble hummingbirds.

DAVID LEGROS

WATER ARUM

Calla palustris
Arum family (Araceae)

Description: Perennial, 5"–12" tall, with jointed stems. Broad, glossy, heart-shaped leaves with pointed tips are 2¼"–5" long and arise on stems 4"–8" long. A broad white spathe surrounds a 1½"- to 2½"-long elongated spadix of tiny, yellowish flowers. Fruit is a red berry when mature.

Bloom season: Late spring to midsummer

Range/habitat: Northern portions of North America in bogs, ponds, wetlands, slow-moving streams, and swamps. The plant may be aquatic or terrestrial-emergent.

Comments: *Calla* may be from the Greek word *kallas* ("beautiful") in reference to the flowers. *Palustris* ("of marshes" or "marsh-loving") refers to the habitat where the plants are found. The foliage contains calcium oxalates, which are toxic compounds; however, the roots are edible if extensively processed. In parts of its range, black bears have been known to eat the young leaves if food is scarce. Also known as wild calla, bog arum, or water-dragon.

BORREGOWILDFLOWERS.ORG

DESERT-LILY

Hesperocallis undulata
Asparagus family (Asparagaceae)

Description: Perennial; stout flowering stems are 1'–3' tall (6' tall in rare instances). Several strap-like leaves arise from the top of the bulb; these basal leaves may be up to 20" long, linear to lance-shaped, bluish-green, and have wavy margins. Trumpetlike flowers are borne along an elongated flowering stalk and have 6 petals with a gray-green or greenish-yellow stripe down the middle of the undersides. Fruit is a capsule.

Bloom season: Late winter through spring

Range/habitat: Southwestern United States and northwestern Mexico in sandy soils in Mohave or Sonoran Desert communities

Comments: *Hesperocallis* is from the Greek *hesperos* ("west") and *kallos* ("beauty"), which translates to "western beauty" and refers to the spectacular beauty of these flowers. *Undulata* ("wavy") refers to the margins of the leaves. The first scientific specimen was collected by J. G. Cooper (1830–1902), an American doctor and naturalist who collected the first scientific specimen. The garlic-flavored bulbs were consumed raw, boiled, or baked by Indigenous peoples in the Southwest. Also known as ajo lily; *ajo* is Spanish for "garlic."

SAND LILY

Leucocrinum montanum
Asparagus family (Asparagaceae)

Description: Perennial, from deeply buried roots. Numerous strap-like leaves arise from the top of the root crown, as these plants lack stems. Tufted leaves are up to 10" long and have a whitish margin. Fragrant flowers are borne on stalks, 2"–5" long, that may barely rise above the ground. Showy, white flowers are often borne in clusters; individual flowers are 1"–2" across. Fruit is an egg-shaped capsule.

Bloom season: Mid-spring to early summer

Range/habitat: Western United States in sandy or rocky sites in sagebrush steppe, prairies, and open woodlands

Comments: *Leucocrinum* is from the Greek *leukos* ("white") and *crinon* ("lily") and refers to the color and type of flowers. *Montanum* ("of the mountains") refers to the plant's distribution. Plants may lie dormant during dry years and then sprout the following spring when soil moisture is sufficient. An early-season bloomer, once the fruits form the plants wither, leaving only the maturing capsules. How the seeds are dispersed and which insects pollinate the flowers are two aspects of these flowers that needs more investigation, as there is scant research on these topics. Also known as star-lily.

STEVE R. TURNER

CANADIAN MAYFLOWER

Maianthemum canadense
Asparagus family (Asparagaceae)

Description: Perennial, low growing, 3"–10" tall. The 1–3 oval-shaped leaves are 3" long and 2" wide and slightly heart-shaped or notched at the base. Leaves may or may not have a stem up to 2" long. Flowering stem often has a slight zigzag pattern and bears 12–25 flowers in a dense flowering cluster; flowers are 1"–2½" long. White starlike flowers are ¼" wide and have 2 sepals, 2 petals, and 4 stamens. Fruit is a reddish berry about ¼" wide.

Bloom season: Late spring to midsummer

Range/habitat: Northern North America to the southeastern United States in wet boggy areas in coniferous, deciduous, or mixed woodlands

Comments: *Maianthemum* is from the Greek *maios* ("May") and *anthemon* ("flower"), referring to the plant's springtime flowering period. *Canadense* ("of Canada") indicates the plant's distribution. The flower's petals and sepals are sometimes referred to as 4 tepals. The plants may spread by rhizomes or creeping roots, often blanketing an area. Plants with 1 leaf are sterile; those with 2 (sometimes 3) leaves produce fertile flowers. Also known as two-leaved Solomon's seal, Canadian May-lily, or Canada beadruby, for the fruit, which wildlife such as ruffed grouse, white-footed mice, and snowshoe hares consume.

STEVE R. TURNER

FEATHERY FALSE SOLOMON'S SEAL

Maianthemum racemosum
Asparagus family (Asparagaceae)

Description: Perennial; often growing in clumps with upright or arching stems, 1'–3' tall. Broad, egg-shaped leaves are 3"–10" long with parallel venation. Numerous tiny white flowers are borne in a terminal cluster with side branching. Individual flowers are tiny, with 6 distinctive tepals. Fruits are fleshy, red berries that may have purplish spots.

Bloom season: Spring to early summer

Range/habitat: Widely distributed across much of North America and northern Mexico in moist meadows, along stream banks, and in woods

Comments: *Maianthemum* is from the Greek *maios* ("May") and *anthemon* ("flower"), referring to the plant's springtime flowering period. *Racemosa* ("flowers in a raceme") describes the floral arrangement. Native Northwest tribes used the boiled roots as either a tea for rheumatism, back injuries, or kidney problems or a poultice for wounds. Although the berries are nontoxic and eaten by wildlife, they are poor tasting.

STEVE R. TURNER

SMOOTH SOLOMON'S SEAL

Polygonatum biflorum
Asparagus family (Asparagaceae)

Description: Perennial, 1'–5' tall; generally with unbranched, arching or erect stems. Jointed stems show a zigzag pattern. Alternate oval to egg-shaped leaves are up to 4" long, with parallel venation, and are smooth on both sides. Tube- or bell-shaped flowers are greenish yellow or greenish white and often arise in pairs and hang downward beneath the arching stem. The ½"- to 1"-long flowers have 6 flaring lobes. Fruit is a ¼"-long blue-black berry.

Bloom season: Mid-spring to early summer

Range/habitat: Central and eastern North America in moist or dry rocky thickets or woods

Comments: *Polygonatum* is from the Greek *poly* ("many") and *gonu* ("knee joint"), referring to the rhizomes, which have many joints. *Biflorum* ("having twin flowers") refers to the pairs of flowers, although the flower groups may have more or less than 2 flowers. Native peoples harvested the potato-like rhizomes as a food source, and early medical practitioners believed that plants with jointed roots were beneficial for treating achy joints and arthritis and as an anti-inflammatory. Where the stalk breaks away from the rhizome, the scar supposedly resembles the Hebrew King Solomon's official seal; hence the common name.

KEN KNEIDEL

SPANISH BAYONET

Yucca baccata

Asparagus family (Asparagaceae)

Description: Perennial. Dense basal cluster of stout, swordlike leaves 12"–40" long and 1"–2" wide. Short stems are single or clumped together. Fibers curl up along the edges of the leaves. Flowering stalks may barely rise above the leaves and bear bell-shaped cream to white flowers that are 1½"–3½" long. Fruits are large and fleshy capsules at maturity.

Bloom season: Mid-spring to early summer

Range/habitat: Southwestern United States and northern Mexico in desert woodlands, shrublands, and canyons

Comments: *Yucca* is the Carib name for manihot, or cassava, although the plants don't resemble each other. *Baccata* ("fruited") refers to the large pods, which, like the young flowering stalks, are edible. Native Americans teased the leaf fibers apart to make cordage, which was used for making sandals, mats, baskets, and cloth. A sudsy soap was made from mixing water with pounded roots and agitated to create foam. Species in the *Yucca* genus have a unique symbiotic pollination relationship with yucca moths (*Tegeticula* sp.): The female moth transfers pollen from one flower to another and deposits her eggs in the flower's ovary, where the larvae will feed on the developing seeds. Also known as broad-leaved yucca.

ADAM'S NEEDLE

Yucca filamentosa

Asparagus family (Asparagaceae)

Description: Perennial, 4'–8' tall. Stout, swordlike, blue-green leaves are up to 30" long and 4" wide, with long curling leaf fibers along the edges. From the basal rosette of leaves, which may be 2'–3' tall, arises a long flowering stalk, often 5'–10' tall (sometimes to 12'), bearing a dense cluster of white flowers. The bell-shaped flowers have 6 tepals and 6 stamens, are 2"-2½" across, and drooping. Fruit is a capsule up to 2" long.

Bloom season: Mid-spring to late summer

Range/habitat: Southeastern United States in sand dunes, beaches, barren areas, and fields

Comments: *Yucca* is from the Carib name for manihot, or cassava, and though the plants don't resemble each other, they do share the trait of enlarged root. *Filamentosa* ("filament" or "thread") refers to the long, curling leaf fibers. Similar to many *Yucca* species, the flowers and fruits are edible. The Cherokee spread powdered roots on the water to "intoxicate" fish, which were then easily captured. Also known as needle-palm, spoon-leaf yucca, or common yucca.

YARROW

Achillea millefolium
Aster family (Asteraceae)

Description: Perennial; flowering stalks up to several feet tall. Fernlike basal leaves are pinnately divided; lower leaves are stalked but smaller, and upper leaves lack a stalk. Flat-topped clusters of small white flowers contain small, ¼"-wide heads with tiny disk flowers and several ray flowers (3–5); rays are whitish or rarely pinkish. Fruit is a tiny black seed without any appendages.

Bloom season: Mid-spring to fall

Range/habitat: Widely distributed across North America in grasslands, meadows, shrublands, and disturbed areas.

Comments: *Achillea* is named for the mythological Greek hero Achilles. *Millefolium* ("1,000-leaved") refers to the finely divided leaf segments. Crushed foliage has a unique odor. In medieval times, yarrow was used as a poultice to stop bloody wounds; hence another common name: soldier's woundwort. This is a somewhat variable plant and widely used in horticultural settings, where it attracts butterflies and a host of insect pollinators.

PEARLY EVERLASTING

Anaphalis margaritacea
Aster family (Asteraceae)

Description: Perennial, with underground stems (rhizomes) bearing numerous upright, unbranched stems covered with white hairs. These woolly stems may grow 1'–4' tall. Numerous lance-shaped or linear leaves bear a prominent midvein and are white-woolly below and green above. Small yellow disk flowers are borne in dense clusters, about ¼" wide, and are surrounded by white involucre bracts, which resemble petals, with a dark basal spot. Fruit is a tiny seed that may bear some white pappus hairs.

Bloom season: Summer through fall

Range/habitat: Widespread across North America in forests, meadows, rocky slopes, roadsides, and pastures

Comments: *Anaphalis* is a near anagram of *Gnaphalium*, a similar-looking genus. *Margaritacea* ("pearl-like") identifies the shape and color of the flower heads. The dry white involucre bracts, which make the flower heads look round, retain this pearly coloration when dried; hence the common name. Native Americans used these plants to treat sores or burns, rheumatism, and colds and as a tobacco substitute. The flowers may be either totally male or female, with a few male flowers to ensure cross-pollination and reduce the risk of self-fertilization. Moths and butterflies are common pollinators for these plants.

SAM KIESCHNICK

WESTERN DAISY

Astranthium ciliatum
Aster family (Asteraceae)

Description: Annual; stems often unbranched, with 3–4 fine ridges, and up to 20" tall. Oval to spatula-shaped leaves form a basal rosette several inches wide. Upper sides of the leaves have long hairs. Lower leaves are up to 1½" long; upper stem leaves are smaller and narrower. Flower heads are borne on stems, 1"–3" long; as the heads mature, they go from upright to drooping then back to upright at maturity. The ½"- to 1"-wide flowering heads have 13–24 ray flowers surrounding a dome-like cluster of yellowish disk flowers. Ray flower straps are white to lavender and ¼"–½" long. About 20 lance-shaped bracts subtend the flower heads in a single row.

Bloom season: Mid-spring to midsummer

Range/habitat: Central United States into northern Mexico in grasslands, meadows, and open areas in woodlands

Comments: *Astranthium* is from a Greek word meaning "starlike," for the shape of the flowers. *Ciliatum* ("fringed with hairs") refers to the plant's pubescence. Also known as Comanche western daisy.

BORREGOWILDFLOWERS.ORG

WHITE TACK-STEM

Calycoseris wrightii
Aster family (Asteraceae)

Description: Annual, generally 2"–12" tall but may reach up to 18" tall. Stems have small, tack-shaped glands and a milky sap. Pinnately divided leaves have linear, threadlike lobes, are up to 5" long and alternately arranged. White flower head is 1½"–2" wide; the strap-like ray flower is 5-lobed, square at the tip, and has 2 reddish veins on the underside. The bracts that subtend the flower head have small glandular hairs and are arranged in 2 rows; the upper row has linear bracts about ½" long, and the lower row has smaller ones that curve backward. Fruit is a brownish seed with fine white hairs.

Bloom season: Spring

Range/habitat: Southwestern United States and northern Mexico in open areas, desert washes, rocky slopes, and shrublands

Comments: *Calycoseris* is from the Greek *kalux* ("cup") and *seris* ("Chicory-like genus"), for the flower's resemblance to those of chicory. *Wrightii* honors Charles Wright (1811–1885), an American botanist and plant collector who served on the US–Mexican Boundary Survey (1851–1852) and collected plants in Texas (1837–1852), Cuba, and Connecticut. The common name refers to the small, tack-like glands along the upper portions of the stems.

JIM FOWLER

WOOLLY SUNBONNETS

Chaptalia tomentosa
Aster family (Asteraceae)

Description: Perennial; leaves arranged in a basal rosette. Each leaf is elliptical, up to 10" long (mostly 4"–6"), and densely hairy below. Flower heads are borne on long stems, 6"–8" long, with creamy white ray flowers that have a purplish underside surrounding a center of white disk flowers. Fruit is a seed with a long, slender, hairy neck.

Bloom season: Late winter to spring

Range/habitat: Southern United States in bogs, savannas, seeps, and wet pinewoods

Comments: *Chaptalia* honors Jean-Antoine Chaptal (1756–1832), a French chemist and industrialist who introduced reforms in medicine and public works under Napoleon I. *Tomentosa* ("with dense hairs") refers to the leaves and flowering stems. Also known as pineland daisy and night-nodding bog-dandelion.

KAREN E. ORSO

ELK THISTLE

Cirsium scariosum
Aster family (Asteraceae)

Description: Perennial; highly variable, including basal rosette form, mound form, and tall form. When a stem is present, it is generally woolly and ridged. Large leaves, up to 16" long, have spiny edges and are either sharply toothed along the margins or cut into deeply toothed lobes. Strap-like leaves are densely clustered along the stem. Flower heads are 1½"–2" wide and consist of smaller flower heads tightly packed together. A row of spiny or toothed bracts subtend the heads and resemble feathery bristles. Individual heads bear white to reddish or purple disk flowers that fade to brown with age. Fruit is a seed with long hairs.

Bloom season: Summer

Range/habitat: Western North America and scattered populations in eastern Canada in meadows and woodlands

Comments: *Cirsium* is from the Greek *kirsion* ("swollen vein") and refers to a species in this genus that was used to treat swollen veins. *Scariosum* ("thin, dry, or translucent") may refer to the light-colored leaves or the bract tips that subtend the flower heads. This plant may be in one of three forms: a low-growing basal rosette with central flower, a clump or mound form with a short stem, or a tall version that reaches about 80". Also known as meadow thistle.

WAVY-LEAF THISTLE

Cirsium undulatum
Aster family (Asteraceae)

Description: Perennial, 2'–5' tall. Stems are covered with dense white hairs. Leaves form basal rosettes; leaves are 2"–12" long and are deeply divided or pinnately lobed. Lobes are toothed or lobed, with yellowish spines along the margins. Leaves often have a coating of fine hairs, making the leaves appear gray. Urn-shaped flower heads, 1½"–2½" wide, are creamy white or pink and contain only disk flowers. Each flower head is a compilation of individual flowers. Bracts below the heads are 1"–2" long, lance-shaped, and have a sticky ridge down the middle; the outer ones have spiny, spreading tips. Fruit is a seed with light hairs attached.

Bloom season: Mid-spring to midsummer

Range/habitat: Widespread across western and central North America in disturbed sites, prairies, shrublands, and woodlands

Comments: *Cirsium* is from the Greek *kirsion* ("swollen vein"), which refers to one species of the genus that was used to treat swollen veins. *Undulatum* ("undulating") refers to the wavy margin of the leaves. Native Americans used the roots for food and as an eyewash to treat infections. Another common name is gray thistle for the color of the leaves.

STEVE R. TURNER

FALSE DAISY

Eclipta prostrata
Aster family (Asteraceae)

Description: Annual; purplish stems upright or prostrate, up to 3' long. Linear leaves are opposite (mostly), up to 5" long and 1" wide. Leaves have some appressed hairs on the upper side and a few teeth along the margins. The ⅓"-wide pie-shaped flower heads arise from the leaf axils and have 8–16 triangular-shaped bracts that subtend the heads. Numerous whitish ray flowers encircle a center of whitish disk flowers. Fruit is a seed with 2–4 teeth on the top.

Bloom season: Midsummer to fall, but may bloom any month of the year

Range/habitat: Widely distributed across the southern United States and eastern North America in moist places such as along ponds, rivers, meadows, wetlands, or ditches

Comments: *Eclipta* is from the Greek *ekleipta* ("to be deficient"), referring to the lack of a pappus on the seeds. *Prostrata* ("prostrate") refers to the prostrate stems, which may form roots at the nodes. Used in herbal medicines to treat liver issues and hair loss. Also known as pie plant or tattoo plant, for the blue ink made from the plant's juice and used for tattoos.

ELEANOR DIETRICH

ELEPHANT'S FOOT

Elephantopus elatus
Aster family (Asteraceae)

Description: Perennial; hairy stems up to 1'–4' tall. Leaves are mostly basal, lance-shaped to elliptical, and 8"–10" long and 1"–3" wide. Stem leaves smaller and scalloped or with teeth along the margins. Flower heads are in small, tight clusters that lack ray flowers; flowers are white to lavender. Heads are subtended by 3 hairy, delta-shaped leafy bracts. Fruit is a seed.

Bloom season: Mid-spring to mid-fall

Range/habitat: Southeastern United States in dry woods, pinelands, sandhills, or oak hammocks

Comments: *Elephantopus* is from the Greek *elephantos* ("elephant") and *pous* ("foot") and refers to the outline of the leaves of a related species resembling an elephant's footprint. (You have to use your imagination a little.) *Elatus* ("raised" or "tall") may refer to the leaves. The plants contain elephantin and elephantopin—two germicidal compounds used to treat respiratory ailments.

STEVE R. TURNER

PHILADELPHIA FLEABANE

Erigeron philadelphicus
Aster family (Asteraceae)

Description: Biennial or short-lived perennial, up to 3' tall (may reach 5'). Stems have varying degrees of white appressed hairs. Spoon-shaped basal leaves often wither before the plant flowers. Egg- to inversely lance-shaped leaves are alternate, up to 4" long, clasp the stem at the base, are pointed at the tip, and often have fine hairs and teeth along the margins. Flower heads arise from leaf axils or the terminal end of the stem, are ½"–¾" wide, and consist of numerous (100–300) white to pinkish narrow ray flowers surrounding a center of yellow disk flowers. One to 3 rows of numerous bracts, which may or may not have hairs and minute glands, subtend the flower heads. Fruit is a seed with light brown hairs.

Bloom season: Late spring to midsummer

Range/habitat: Widely distributed over much of North America in moist fields, prairies, meadows, river edges, disturbed ground, or along shallow ponds

Comments: *Erigeron* is from the Greek *eri* ("early") and *geron* ("an old man"), referring to the white flower heads. *Philadelphicus* ("of Philadelphia") refers to where the plant was first collected for science. Across its range, this fleabane is known as marsh fleabane or poor robin's plantain.

JUDY PERKINS

WHITE TIDY-TIPS

Layia glandulosa
Aster family (Asteraceae)

Description: Annual, 4"–24" tall, with thin greenish-purple stems. Narrow basal leaves are lobed or toothed, ½"–3" long, and covered with stiff, glandular hairs; upper leaves are not lobed. Flowering heads are up to 1½" wide and have a center of yellow disk flowers surrounded by several (3–15) white, 3-lobed (toothed) ray flowers. Bracts subtending the flowering heads may have sticky glands. Fruit is a seed with about 10 flat, white hairs.

Bloom season: Spring

Range/habitat: Western North America in open areas, fields, meadows, grasslands, and shrublands

Comments: *Layia* honors George Tradescant Lay (1799–1845), an English naturalist on the Beechey voyage (1825–1828) who collected plants in Asia, Hawaii, Alaska, California, and South America. *Glandulosa* ("glandular") refers to the sticky hairs that cover the plant. The common name refers to the "tidiness" of the ray flowers. During moist springs this plant will bloom in profusion. Ground-foraging birds and rodents consume the edible seeds, which some Native tribes collected and ground into flour. The plants may give off a spicy scent.

MOHAVE DESERT-STAR

Monoptilon bellioides
Aster family (Asteraceae)

Description: Annual, ½"–2" tall; in wet years may reach 10" tall. Reddish-purple stems are covered with stiff hairs, mostly prostrate, and spreading just above the ground. Linear leaves are ⅓"–¾" long, succulent-like, covered with stiff hairs, and have a blunt tip. The ¾"-wide flowering heads have several (6–20) white to pale pink ray flowers surrounding a center of yellow disk flowers. Flower heads seem large in contrast to the rest of the plant. Fruit is a seed with bristly hairs.

Bloom season: Late winter through spring; may flower in late summer, depending on summer rainfall

Range/habitat: Southwestern United States and northwestern Mexico in open, stony ground, desert washes, and rocky slopes

Comments: *Monoptilon* is from the Greek *monos* ("one") and *ptilion* ("feather"), referring to the seed's bristlelike pappus. *Bellioides* ("like *Bellis*") refers to the flowers resembling those in the *Bellis* genus; *bellis* means "pretty." Above-average winter moisture in the Mohave or Sonoran Desert will affect the blooming period and growth of these plants and may result in the flowers carpeting the desert. The flowers attract butterflies, moths, and other insects as pollinators. Also known as desert-star or bristly desertstar.

CHUCK TAGUE

TALL RATTLESNAKE ROOT

Nabalus altissimus
Aster family (Asteraceae)

Description: Perennial, 2'–6' (occasionally 7') tall, mostly branching near the top of the plant. Stems and leaves exude a milky sap when broken. Leaves are arranged alternately along the stem and margins. Largest leaves may be 8" long and 6" wide. Pendulous flowering heads arise on short stems, often branching off the main stalk. The ¾"-wide flowering heads lack disk flowers and are made up of 5–6 tubular ray flowers. Flowering heads are subtended by 5 bracts that are light green, hairless, and about ½" long; smaller secondary bracts are at the base of these primary bracts. Fruit is a seed with a small tuft of tan or orange-brown hairs.

Bloom season: Late summer to fall

Range/habitat: Eastern North America to Texas and Florida in sandy to rocky woods

Comments: *Nabalus* refers to a town in Palestine where a related species is from. *Altissimus* ("very tall") indicates the height of the plant. The common name is in reference to the Iroquois making a poultice from the roots to treat rattlesnake bites. Also known as tall wirelettuce. White rattlesnake-root (*N. albus*) is another "white lettuce" that grows in eastern North America and has a reddish-brown pappus.

SWEET COLTSFOOT

Petasites frigidus
Aster family (Asteraceae)

Description: Perennial. From slender rhizomes arise numerous flowering stems, 5"–25" tall, before the leaves appear. Stem leaves are small, but the deeply divided basal leaves may be 1' wide and heart- or kidney-shaped, with 5–7 toothed lobes. Upper sides of the leaves are green and hairless; undersides are woolly. Flower heads are arranged in flat-topped clusters, which may appear rounded at first; individual heads are ½"–¾" wide. Creamy white to pink ray flowers surround a cluster of white or pinkish disk flowers. The lance-shaped bracts that subtend the flower heads have hairy bases. Plants may have only male or female flowers or both. The ribbed seeds bear a crown of numerous white hairs.

Bloom season: Spring

Range/habitat: Across northern North America and from the Pacific Northwest to the northeastern United States in moist forests, swamps, meadows, and along roadsides and lake edges

Comments: *Petasites* is from the Greek *petasos* ("a broad-brimmed hat") and refers to the large basal leaves. *Frigidus* ("growing in cold regions") refers to the plant' distribution. The common name comes from a European relative with leaves that resemble the shape of a young horse's hoof. The light, winged seeds are easily dispersed by the wind, enabling this plant to colonize newly disturbed areas. The leaf stalks and flowering stalks are edible. Also known as palmate coltsfoot or Arctic sweet coltsfoot.

JUDY PERKINS

NEW MEXICO PLUMESEED

Rafinesquia neomexicana
Aster family (Asteraceae)

Description: Annual, 6"–20" tall, with smooth purplish stems that zigzag and have milky sap. Narrow leaves are up to 8" long, with deep lobes that may be up to ¾" long; basal leaves are larger than the upper leaves. The 1½"- to 2"-wide flower heads are subtended by linear-shaped bracts. White, strap-shaped ray flowers are notched at the tip, and the outer flowers are longer than the inner ones; undersides of the larger petals may have a purplish tinge. Fruit is a seed with feathery bristles.

Bloom season: Spring

Range/habitat: Southwestern United States from California to west Texas and south into Mexico in dry washes and on mesa tops and gravelly desert slopes

Comments: *Rafinesquia* honors Constantine Samuel Rafinesque-Schmaltz (1783–1840), a nineteenth-century botanist. *Neomexicana* ("New Mexico") refers to the plant's type locality. The plants often grow in close proximity to shrubs; the weak stems of this plumeseed grow up through the shrub's branches, which provide support. The feathery seed bristles give this plant its common name.

HOARY GROUND DAISY

Townsendia incana
Aster family (Asteraceae)

Description: Short-lived, low-growing perennial with stems ¾"–4" tall. Stems and leaves are covered with conspicuous hairs. Spatula- to inversely lance-shaped leaves are ¼"–1½" long. Flower heads on short stalks are solitary or few, ½"–1" wide, and with 13–34 ray flowers that are white on the upper surface and pink to lavender below. Ray flowers surround a center of yellowish disk flowers. Fruit is a seed with hairs.

Bloom season: Mid-spring to midsummer

Range/habitat: Western United States in shrublands, open areas, and woodlands

Comments: *Townsendia* honors David Townsend (1787–1858), an amateur botanist from West Chester, Pennsylvania, who traded plant specimens from the West Chester area with English botanist William Jackson Hooker for British natural history books. *Incana* ("hoary") refers to the dense white hairs on the stems. Various species of *Townsendia* are difficult to distinguish from one another due to hybridization. Also known as silvery townsendia.

WHITE MULE'S EARS

Wyethia helianthoides
Aster family (Asteraceae)

Description: Perennial, 1'–2½' tall, often covering large areas. Basal leaves are egg-shaped, up to 1' long, and have smooth margins; stem leaves are smaller. Leaf stems are hairy. Solitary, large white flower heads have 13–21, 1"- to 2"-long ray flowers that surround a center of yellow disk flowers. Flowers may turn pale yellow with age. Fruit is a seed.

Bloom season: Late spring to early summer

Range/habitat: Western United States from Oregon to Wyoming in moist meadows, stream banks, and open areas

Comments: *Wyethia* is for Nathaniel Wyeth (1802–1856), the "Cambridge Iceman," who led two expeditions to the Oregon Territory, in 1832 and 1834. *Helianthoides* ("resembling *Helianthus*") refers to the similarity of this plant to sunflowers in the *Helianthus* genus. The showy flowers attract myriad pollinators, including butterflies, bumblebees, flies, and hawkmoths. Also known as whitehead mule's ears, referring to the large, earlike shape of the leaves.

ALEX HEYMAN

MOJAVE WOODY ASTER

Xylorhiza tortifolia
Aster family (Asteraceae)

Description: Perennial, with glandular stems 24"–31" tall. Oval, linear, or lance-shaped leaves also have hairy, glandular surfaces; ends of the leaves are pointed or with spiny tips. Leaf margins have several teeth or spines, and the leaves are up to 4" long. Solitary flower heads arise on long stalks and bear over 60 white to lavender or blue, narrow ray flowers surrounding a center of disk flowers. Flower heads are 1¼"–1¾" wide. Flower head is subtended by a row of narrow bracts, about ¾" long. Fruit is a seed with bristles.

Bloom season: Spring to early summer; may bloom again in the fall if there is sufficient summer rain

Range/habitat: Southwestern United States and northern Mexico in desert washes, rocky slopes, shrublands, and Joshua tree woodlands

Comments: *Xylorhiza* ("woody base") refers to the woody nature of the stem's bases. *Tortifolia* is from *tortus* ("twist") and *folium* ("a leaf"), in reference to the twisted appearance of the leaves. Big Bend aster (*X. wrightii*) occurs in Texas and northern Mexico. The Havasupai carried the leaves in their pockets to mask body odors. Larvae of the sagebrush checkerspot butterfly (*Chlosyne acastus*) feed on the leaves of this aster.

MICHAEL PLAGENS

DESERT ZINNIA

Zinnia acerosa
Aster family (Asteraceae)

Description: Perennial, 4"–10" tall, with woody bases. Narrow linear to needlelike leaves are ¾"–1½" long and arranged oppositely along the stems. Flowering heads have 4–7 whitish ray flowers, which are somewhat toothed at the tip and papery to the touch. Center of the flower head is composed of 8–13 yellow or purplish disk flowers. Flower heads are about 1" wide. Fruit is a seed.

Bloom season: Mid-spring to fall, depending on summer rainfall

Range/habitat: Southwestern United States and northern Mexico in dry canyons, mesas, rocky slopes, and other open areas

Comments: *Zinnia* honors Johann Gottfried Zinn (1727–1759), a German ophthalmologist and botanist who was a professor of medicine and the botanical director in Gottingen, Germany. *Acerosa* ("needlelike") refers to the shape of the leaves. Zinn may be best known for the first accurate illustration of the eye and associated nerves and vessels. He collected Zinnia (*Z. elegans*) seeds, ancestors of the common garden zinnia, while traveling in Mexico. The flowers attract butterflies and other insects as pollinators, and the plant is often used in landscape reclamation projects. A poultice of the crushed leaves has been used to alleviate swelling. Also known as southern zinnia or dwarf zinnia.

VANILLA LEAF

Achlys triphylla
Barberry family (Berberidaceae)

Description: Perennial. From thin, underground rhizomes arise leaf stalks that are 4"–16" tall. The broad leaf is divided into 3 fan-shaped leaflets. Asymmetrical segments are coarsely toothed or scalloped along the margins; the smaller middle segment may be roughly divided into 3 lobes. The tiny, white flowers are borne on a leafless stalk and arranged in a tight, 1"- to 3"-long spike that rises above the leaves. Flowers lack sepals and petals but bear 8–20 white stamens. Small, crescent-shaped, reddish-purple fruits are covered with fine hairs.

Bloom season: Late spring to midsummer

Range/habitat: Western North America in shady forests and along stream banks

Comments: *Achlys* is from the Greek *achlus* ("mist") and is perhaps a reference to the misty appearance of the white flowers. *Triphylla* ("three-leaved") refers to the 1 leaf divided into 3s. The dried leaves have a vanilla-like aroma when crushed. To some, the lobed middle leaf resembles a deer's hoof when bent backward; hence another common name for the plant: deer-foot. Also known as sweet-after-death, in reference to the fragrance of the crushed leaves.

ELEANOR RAY

TWINLEAF

Jeffersonia diphylla
Barberry family (Berberidaceae)

Description: Perennial, 8"–18" tall. At the top of the leaf stem, the blue-green leaf, about 6" long, is divided into 2 equal segments. The 1"-wide, cup-shaped white flowers have 8 petals and arise on a leafless flowering stalk. The 4 sepals drop off early. Fruit is a pear-shaped capsule with a lid.

Bloom season: Spring

Range/habitat: Eastern North America in limestone soils in moist woodlands

Comments: *Jeffersonia* honors Thomas Jefferson (1743–1826), third president of the United States and a strong supporter of botany and agriculture. *Diphylla* ("two leaves or leaflets") describes the arrangement of the leaves and the common name. The leaves are about half-tall (8"–12") when the flowering stalk appears. The leaves continue to grow after the flowers appear. Native peoples made a tea from the roots to treat intestinal disorders and urinary ailments. The plants contain the alkaloid berberine, which is considered toxic. Another common name is rheumatism root.

ELEANOR DIETRICH

MAYAPPLE

Podophyllum peltatum
Barberry family (Berberidaceae)

Description: Perennial, up to 1'–1½' tall. As the leaves first appear, they resemble a closed umbrella, then open to reveal 1 or 2 large, toothed leaves that are 6"–12" wide. If there are 2 leaves, they are arranged opposite each other. Apple-like, white to rose flowers are 1½"–3" wide and have 6–9 waxy petals and numerous stamens. Nodding flowers arise on a short stem from the same node where the leaf stems emerge. Fruit is a lemon-shaped berry.

Bloom season: Late spring to early summer

Range/habitat: Eastern North America in open, damp woodlands

Comments: *Podophyllum* is from the Greek *podos* or *pous* ("foot") and *phyllon* ("leaf"), in reference to the shape of the leaf; another common name for the plant is duck's foot. *Peltatum* ("peltate leaves" or "shield shape") refers to the more or less rounded leaf with the stem attached at a central point on the underside of the leaf. The bright yellow and slightly soft edible fruit should only be eaten when ripe; unripe fruit may cause digestive ailments. Plants with a single leaf do not produce a flower. The leaves, roots, and seeds contain podophyllin, a toxic compound. The common name refers to the May blooming of the apple-like flower. Also known as mandrake.

KEN-ICHI UEDA

NORTHERN INSIDE-OUT FLOWER

Vancouveria hexandra
Barberry family (Berberidaceae)

Description: Perennial, up to 20" tall. Compound leaves arise on long basal stalks and are divided into 3 divisions that bear 9–15 heart- to egg-shaped leaflets. The flowering stalks bear small white, starlike, nodding flowers. Flowers have 6 sepals and petals; petals are shorter than the sepals and have hooded tips. Both the sepals and petals bend backward and flare open at the base. Fruit is a purplish pod with sticky hairs that bears black seeds.

Bloom season: Late spring to early summer

Range/habitat: Western North America in shady locations in coniferous forests

Comments: *Vancouveria* is for Capt. George Vancouver (1757–1798), the British explorer who sailed twice with Captain Cook and who explored and mapped the Pacific Northwest coast from 1791 to 1795. *Hexandra* ("six stamens") refers to the number of stamens. The flowers have 6–9 outer sepals that fall off before the flower opens. The black seeds have a fleshy appendage that almost covers the seed; this coating attracts ants and wasps, which help disperse the seeds.

CHRYSANTHEMUM-LEAF EUCRYPTA

Eucrypta chrysanthemifolia
Borage family (Boraginaceae)

Description: Annual, up to 3' tall; stems generally about 1½' long and with glandular hairs. Oval-shaped leaves have a lacy appearance, as they are lobed; these lobes may be further divided into smaller lobes. Leaves are ¾"–5" long; upper leaves have fewer divisions. Flowers are borne in small groups, 4–15 per stem, ⅛"–¼" wide, and an open bell shape. The white petals have fine purplish nectar guide lines leading into the flower's throat. Fruit is a bristly capsule with two types of seeds: larger ones that are wrinkled and asymmetrical in shape and others that are smoother, smaller, and "hidden" behind the larger ones.

Bloom season: Midwinter to early summer

Range/habitat: Southwestern United States into northern Mexico in coastal sage shrub, chaparral, and oak woodlands

Comments: *Eucrypta* is from the Greek *eu* ("well or true") and *crypta* ("secret"), referring to the smaller seeds hidden behind the larger ones. *Chrysanthemifolia* ("with *Chrysanthemum*-like leaves") refers to the resemblance of these leaves to those on a *Chrysanthemum* species plant; hence the common name as well. These aromatic plants have a "vinegary" odor and are also known as spotted hideseed, for the larger seeds that "hide" the smaller seeds.

BORREGOWILDFLOWERS.ORG

SALT HELIOTROPE

Heliotropium curassavicum
Borage family (Boraginaceae)

Description: Perennial, growing upright to 2½' or with stems prostrate along the ground. Stems and leaves are fleshy. The ¼"- to 1½"-long leaves are oval to spade-shaped, thick, and blue-green. Bell-shaped white flowers are arranged in dense flowering clusters that are coiled or curved, generally about 4" long but can reach 1' long. The ¼"-long flowers have 5 lobes, a purple to yellowish throat, and are arranged in double rows along a stem. Fruit is a nutlet.

Bloom season: Spring to fall

Range/habitat: Much of North America, south to Argentina, and in Hawaii and the West Indies in salty, sandy areas such as dunes, beaches, and alkali flats

Comments: *Heliotropium* ("sun turning") is for the summer solstice, when the flowering plants were first recorded for science. *Curassavicum* ("of Curacao") refers to the location of the first plants collected for science. The plants produce pyrrolizidine alkaloids, which defend the plant against insect herbivory. Although containing somewhat toxic compounds, the plant has been known to be used as a potherb. The flowers attract a wide variety of butterflies as pollinators. Also known as seaside heliotrope, monkey's tail, or quail plant, for its preferred habitat, the coiled flowering stalk, and birds that consume the seeds, respectively.

QUINTEN WIEGERSMA

MARBLESEED

Onosmodium molle
Borage family (Boraginaceae)

Description: Perennial, 2'–4' tall, with hairy stems. Alternate egg- to lance-shaped leaves are also hairy, up to 5" long, with prominent longitudinal veins and a sharp point. Flowering stems, up to 6" long, bear numerous flowers arranged in a tight coil. Tubular-shaped flowers have 5 white to greenish-white petals and are about ½" long; the style extends beyond the mouth of the flower. Fruit is a nutlet.

Bloom season: Mid-spring to midsummer

Range/habitat: Central North America and portions of the eastern United States in prairies, savannas, pastures, thickets, and open woods

Comments: *Onosmodium* ("resembling *Onosma*") refers to the plants resembling those in the *Onosma* genus. *Molle* ("soft") refers to the texture of the leaves. The hard nutlets give this plant its common name.

STEVE R. TURNER

CANADIAN ROCKCRESS

Boechera canadensis
Mustard family (Brassicaceae)

Description: Biennial; flowering stem is 1'–3' tall. Inversely lance-shaped to shallowly pinnate basal leaves are 2½"–5" long, smooth to moderately hairy, have a short stem, and may be sparsely toothed along the margins. Stem leaves are elliptical to lance-shaped, lack a stalk, and are smaller as they progress upward along the stem. A 4"- to 16"-long flowering stalk arises at the end of the main stems and bears ¼"-long whitish flowers with 4 white petals, 4 green sepals, and 6 stamens; the spreading petals are about half again as long as the sepals. Fruit is a narrow seedpod (silique), 1½"–4" long, that arches downward as it ages.

Bloom season: Late spring to midsummer

Range/habitat: Throughout much of central and eastern North America in sandy sites in savannas, woodlands, shaded cliffs, and slightly wooded dunes or bluffs

Comments: *Boechera* honors Danish botanist Tyge W. Böcher (1903–1983), who worked on the taxonomy of alpine plants. *Canadensis* ("of Canada") refers to where the first plant was collected for science. Small flies and bees are the primary pollinators for these flowers. The seedpods split open at maturity and release small seeds, which have a papery covering to help them be dispersed by the wind. Also known as sicklepod rockcress.

STEVE R. TURNER

WHITE SPRINGCRESS

Cardamine bulbosa
Mustard family (Brassicaceae)

Description: Perennial, ½'–1½' tall; stems mostly unbranched. Basal leaves arise on stems longer than the leaves and have oval, round, or kidney-shaped blades, about 1½" long, with undulating margins. Stem leaves are narrower and longer than the basal leaves and are smooth, alternate, about 2" long and 1" wide, with margins that are undulate, toothed, or entire. The flowering stalk is borne at the end of the stems and bears white, ½"-wide flowers with 4 rounded petals, 4 sepals, and 6 stamens; petals are much longer than the sepals. Fruit is an erect 1"-long seedpod with smooth seeds.

Bloom season: Late spring to late summer

Range/habitat: Central and eastern North America in moist sites such as riverbanks, meadows, vernal pools, woodlands, and shaded cliffs

Comments: *Cardamine* is the ancient Greek name for a type of cress. *Bulbosa* ("bulbous") refers to the swollen roots. Bees, flies, and small butterflies are attracted to the flowers for nectar and pollen. The edible leaves taste peppery, which indicates another common name: bittercress.

SPECTACLE-POD

Dimorphocarpa wislizeni
Mustard family (Brassicaceae)

Description: Annual, with several to many erect stems to 20" tall. Basal leaves are gray, wavy-toothed, or irregularly lobed, 1"–3" long, and covered with minute hairs. Leaves along the flowering stalk are smaller. Flowers are white to greenish, ⅜"–¾" wide, arising in ladderlike fashion, with the top flowers arranged in a dense cluster. Fruit is a rounded seedpod, ½" wide, fused together along a common midline.

Bloom season: Mid-spring to early summer

Range/habitat: Southwestern United States from Nevada to Texas in sandy dunes, desert shrublands, and woodlands

Comments: *Dimorphocarpa* ("two-formed fruit") describes the seedpods, which also resemble a pair of eyeglasses; hence the common name. *Wislizeni* honors Friedrich Adolph Wislizenus (1810–1889), a German physician, traveler, and author who immigrated to the United States in 1835, joined a trading caravan to Mexico in 1846, and made many observations of the local flora and fauna during his travels.

MOUNTAIN PEPPERGRASS

Lepidium montanum
Mustard family (Brassicaceae)

Description: Perennial (less commonly, biennial), growing up to 4' tall; a highly variable species. Leaves are basal only or basal and along the stem, variously shaped, and ¼"–5" long. Basal leaves may be cleft to midline or entire; stem leaves are narrow and entire along the margins. White flowers, 4-petaled and up to ⅛" across, are arranged in tight clusters. There are 6 stamens, 4 tall and 2 short. Seedpods are ⅛" long, flattened, and egg-shaped.

Bloom season: Mid-spring to midsummer

Range/habitat: Western United States in open desert, mixed desert shrubs, grasslands, and woodlands

Comments: *Lepidium* is from the Greek *lepis* ("scale"), a reference to the flattened shape of the seedpods. *Montanum* ("of the mountains") refers to the elevational distribution of this plant. The seedpods are edible and peppery. The form of this plant is highly variable. Fremont's peppergrass (*L. fremontii*) grows from California to Arizona and is similar to mountain peppergrass but has larger white petals and heart-shaped fruits.

PATRICK ALEXANDER

MOUNTAIN CANDYTUFT

Noccaea fendleri
Mustard family (Brassicaceae)

Description: Perennial, 1"–14" tall. May grow in profusion. Leaves are linear, oblong, egg- or spatula-shaped, up to 3" long, green or purplish, and the margins may be toothed. Upper leaves are smaller than the basal leaves. The flowering stalks may be up to 10" long and bear a dense cluster of flowers with 4 white petals, 4 greenish-purple sepals, and 6 stamens; petals are about ½" long. Fruit is a flattened seedpod that is spatula-shaped and pointed at the tip.

Bloom season: Spring and summer

Range/habitat: Western United States in canyons, woodlands, meadows, glades, and alpine areas

Comments: *Noccaea* is for Domenico Nocca (1758–1841), an Italian clergyman and director of the botanical garden of Mantova, Italy. *Fendleri* honors Augustus Fendler (1813–1883), a German immigrant who collected many southwestern plants for Asa Gray, the famous Harvard botanist. This is one of the earliest wildflowers to bloom in the spring.

CUTLEAF THELEPODY

Thelypodium laciniatum
Mustard family (Brassicaceae)

Description: Biennial, from stout stems; upright branches reach 3' or more. Stems are hairless. Basal rosette of leaves is divided into several lance-shaped leaflets or lobes; these leaves wither before flowering. Upper leaves are less divided, smaller, have a stalk, and are pointed at the tip. A long plume of white to pale lavender flowers with linear petals is borne along a central stalk. The linear petals are not wide at the tip. Fruit is a narrow, cylindrical seedpod.

Bloom season: Mid-spring to midsummer

Range/habitat: Western North America in rocky cliffs, plateaus, sagebrush flats, and grasslands

Comments: *Thelypodium* is from the Greek *thelys* ("female") and *podoin* ("little foot"), which refers to the short-stalked ovary of many species. *Laciniatum* ("shredded" or "torn to pieces") refers to the highly dissected leaves. Manyflower thelepody (*T. milleflorum*) also has upright flowering stalks bearing numerous flowers and also grows in the western United States.

JIM FOWLER

ALLEGANY PACHYSANDRA

Pachysandra procumbens
Boxwood family (Buxaceae)

Description: Perennial, 6"–12" tall, but may spread 1'–2' via rhizomes. Stems are fleshy. Blue-green leaves dotted with purple and white marbling are egg-shaped to rounded, coarsely toothed along the upper portion of the blade, and up to 3" long. Across some of the range, the leaves are evergreen. Tiny white to greenish-white flowers are borne in short, 2"- to 4"-long spikes. Male flowers are located above the female flowers. Fruit is a small capsule.

Bloom season: Spring

Range/habitat: Southeastern United States in shady woodlands

Comments: *Pachysandra* is from *pachys* ("thick") and *andros* ("stamen"), referring to the stout stamen's filaments. *Procumbens* ("trailing") refers to the plant's ground-covering habit. The short flowering stalk may be somewhat covered by leaf litter. Asiatic pachysandra (*P. terminalis*) is commonly used in landscaping and tends to overtake areas where it is planted. Also known as mountain pachysandra or Allegheny spurge.

LYNDA PAZNOKAS

SAGUARO

Carnegiea gigantea
Cactus family (Cactaceae)

Description: Large cactus, up to 50'–65' tall, with stout branches. The internal succulent structure consists of numerous wooden rods that form ridges that expand and contract with moisture content. The trunk and branches are covered with spines that may be 3" long. White flowers in dense clusters near the top of the stem are 5" long and 2½" wide and funnel-shaped. Fruit is a fleshy pod.

Bloom season: Mid-spring to early summer

Range/habitat: Southwestern United States from California to Arizona and south into northern Mexico in desert shrublands

Comments: *Carnegiea* honors Andrew Carnegie (1835–1919), an American philanthropist, industrialist, and founder of the Carnegie Institute. *Gigantea* ("gigantic") refers to the stature of these cacti, which are the largest in North America. Saguaro is the state wildflower of Arizona. Long-lived, these cacti may survive to 150–200 years old and won't produce side arms until they are at least 50–75 years old. A saguaro without arms is called a "spear." During times of above-average moisture, these cacti swell up and store excess moisture. Birds, bats, and insects pollinate the flowers.

SIMPSON'S HEDGEHOG CACTUS

Pediocactus simpsonii

Cactus family (Cactaceae)

Description: Solitary or colonial rounded stems grow 1"–7" tall and 1½"–10" wide. Stems are covered with tubercles (swollen areas) that bear brown to blackish central spines and white radial spines. Flowers are ½"–1½" wide; petallike parts range from white to greenish, yellowish, or pinkish. Fleshy fruit is green and may turn reddish with age.

Bloom season: Late spring to midsummer

Range/habitat: Western United States in desert shrub, mountain foothills, and montane woodlands

Comments: *Pediocactus* is from the Greek *pedio* ("plains"), referring to the growing location of this cactus. *Simpsonii* honors James H. Simpson (1813–1883), a topographical engineer who first collected this cactus in Nevada. The growth and flower color are highly variable. Also known as mountain ball cactus, for its elevational range from 4,000' to 11,500'.

PATRICK ALEXANDER

DESERT NIGHT-BLOOMING CEREUS

Peniocereus greggii

Cactus family (Cactaceae)

Description: Perennial cactus. Narrow stems are ½"–1" wide and rise 3'–6' tall from a stout turnip-like tuber. Stems have 4–9 angled ribs and woolly areoles, from which about 12 black or gray spines arise. The beautiful white flowers bloom for 1 night, are about 8"–12" long and 3" wide, with numerous petals and stamens. Flowers release a faint vanilla aroma. Fruit becomes reddish at maturity; birds that eat the fruits disperse the seeds.

Bloom season: Summer

Range/habitat: Southwestern United States and northern Mexico in sandy grasslands and desert shrublands

Comments: *Peniocereus* is from the Greek *pene* ("thread") and Latin *cereus* ("wax taper"), a common name that refers to many different cacti species. Combined, the names refer to this plant's narrow stems. *Greggii* honors Josiah Gregg (1806–1850), an explorer, naturalist, and author who traveled extensively in the southwestern United States and northern Mexico and published his travel notes in *Commerce of the Prairies* in 1844. For most of the year the plant resembles a dead bush and often grows intermixed with other shrubs, which provide some support for the thin stems. Illegal harvesting of the tubers has led to the plant becoming rare in its natural habitat. The flowers bloom in flushes that last for just 1 night. Hawkmoths are the primary pollinators of this plant. Also called Arizona queen of the night and *Reina de la Noche.*

ENRIQUE MARTINEZ NÚÑEZ

ORGAN PIPE CACTUS

Stenocereus thurberi

Cactus family (Cactaceae)

Description: Perennial cactus with multiple upright stems. Stems arise from a common base and are up to 6" wide and 30' tall; the overall plant may be 12' wide. The 12–19 ribs bear numerous 2"-long black spines that turn gray with age. Funnel-shaped white flowers are 2"–3" wide and tinted pink or purple. Fruit is greenish-red and fleshy.

Bloom season: Mid-spring to early summer

Range/habitat: Southern Arizona, Baja California, and northern Mexico in rocky hillsides, desert chapparal, and coastal thorn scrub, and deciduous woodlands

Comments: *Stenocereus* is from the Greek *stenos* ("narrow") and the Latin *cereus* ("wax taper"), referring to the candle- or candelabra-like overall shape. *Thurberi* honors George Thurber (1821–1890), an American naturalist and pharmacist who collected plants in the southwestern United States and northern Mexico during the 1848–1855 US–Mexican Boundary Survey. Bats are the primary pollinators of the night-blooming flowers. The edible fruits, which taste like watermelon, are made into a candy called *pitahaya dulce*. Long-lived, a mature cactus may be around 200 years old.

STEVE R. TURNER

SPIKE LOBELIA

Lobelia spicata

Bellflower family (Campanulaceae)

Description: Short-lived perennial or biennial; slender, unbranched stems rise 1'–3½' tall. Basal leaves are lance- to narrowly spatula-shaped and up to 3½" long. Leaf margins may be slightly toothed or smooth. The 2-lipped white to bluish-white flowers are about ⅓"–½" wide and borne along an elongated stalk that is 3"–12" long. Flower's upper lip is divided into 2 lobes; the lower lip, which has 2 yellowish spots near the throat, is divided into 3 lobes. A dark blue to purplish stigma extends beyond the flower's opening. Fruit is a seedpod.

Bloom season: Summer

Range/habitat: Eastern North America in generally moist savannas, grasslands, prairies, meadows, thickets, river edges, and abandoned fields

Comments: *Lobelia* honors Matthias de l'Obel (anglicized to Matthias de Lobel) (1538–1616), a Belgium botanist and physician to English King James I. *Spicata* ("spike bearing") is in reference to the type of flowering arrangement. When crushed, the leaves and stems exude a milky white latex, which is toxic. Long-tongued bees and small butterflies are common pollinators on these flowers. Native peoples made a tea from the stems as a wash for sore jaws.

TOM LEBSACK

REDWHISKER CLAMMYWEED

Polanisia dodecandra
Caper family (Capparaceae)

Description: Annual or perennial, 1'–3' tall. Leaves and stems are covered with sticky, moist glands. Compound leaves are about 2" long and divided into 3 palmate leaflets. Flowers are borne in dense clusters at the ends of the stems. White to cream flowers are about 1" long, with 4 petals and numerous reddish-purple stamens of unequal length that project beyond the petals. Fruit is a slender seedpod, 1"–2" long, that points upward.

Bloom season: Late spring to fall

Range/habitat: Across much of North America except a few southern states and Maine in sandy or barren sites with little vegetation

Comments: *Polanisia* is from the Greek *polys* ("many") and *anisos* ("unequal") and refers to the numerous stamens, which are of unequal lengths. *Dodecandra* ("having twelve stamens") refers to the number of stamens. The ripening seedpod points upward, which differs from the related spiderweeds (*Cleome* sp.), which point downward. The common name refers to the long red stamens resembling whiskers and the sticky nature of the plant; the sap and leaves have an unpleasant sulfur-like odor. May be assigned to the Cleomaceae family.

TWINFLOWER

Linnaea borealis
Honeysuckle family (Caprifoliaceae)

Description: Perennial; long slender runners arise from leafy stems that are less than 5" tall and 8"–16" long. Leaves are dark green above and broadly elliptical-shaped, with an opposite arrangement. Upper half of the leaf has a few shallow teeth along the margins. A pair of pink, ¼"-long flowers are borne at the end of a Y-shaped stalk. Fruits are small nutlets with sticky hairs.

Bloom season: Summer

Range/habitat: Across much of northern North America from rocky shorelines to subalpine forests

Comments: *Linnaea* is for Carl Linnaeus (1707–1778), a Swedish scientist who devised the current taxonomic binomial system of genus and species. In many portraits of Linnaeus, he is depicted holding a sprig of this plant, his brand for being elevated to Swedish nobility in 1757. *Borealis* ("northern") refers to the plant's distributional range. The common name describes the habit of the fragrant flowers borne in pairs. The flower is also the symbol of Linnaeus's home province of Småland.

JAMIE CARTER

SCOULER'S CATCHFLY

Silene scouleri

Pink family (Caryophyllaceae)

Description: Perennial, with 2'- to 3'-tall, generally unbranched, flowering stalks. Basal leaves have glandular hairs and are inversely lance-shaped at the base and lance-shaped higher up on the stem; leaves vary, 2½"–10" long. Tubular to bell-shaped sepals are fused together and have purplish veins. Petals vary from white to greenish white to rose and have 2 or 4 deep, narrow lobes. Fruit is a capsule.

Bloom season: Summer

Range/habitat: Western North America in grasslands, prairies, woodlands, and rocky bluffs

Comments: *Silene* is probably from the Greek *sialon* ("saliva"), which refers to the stem's sticky hairs, or *selinos*, for Silenus, the intoxicated foster-father of Bacchus (god of wine) who was covered with foam. *Scouleri* honors Dr. John Scouler (1804–1871), a surgeon and naturalist who traveled with David Douglas collecting plants in the Pacific Northwest.

JIM FOWLER

STARRY CAMPION

Silene stellata

Pink family (Caryophyllaceae)

Description: Perennial, 1'–2½' tall; clump forming, with unbranched or sparsely branched stems. Upper and lower leaves are arranged oppositely; middle leaves are in whorls of 4. Leaf nodes are swollen and reddish. Elliptical or lance-shaped leaves are up to 4" long; the upper surface is gray-green and hairless, while the undersides may have some fine hairs. Flowers are arranged in an open cluster about 8" long, singularly or in groups of 2–3. Flowers are ¾" wide and have a green, bell-shaped calyx with 5 triangular-shaped teeth and 5 white, fringed petals. Fruit is a capsule.

Bloom season: Summer

Range/habitat: Central and eastern United States in clearings, meadows, and dry woods

Comments: *Silene* is taken from the Greek woodland god, Silenus. *Stellata* ("starlike") refers to the shape of the petals. Because the flowers open in the evening, hawkmoths are a primary pollinator. Members of this genus are known as catchfly, for the sticky sap on the stems, which traps small insects. Not considered a true carnivorous plant.

REUVEN MARTIN

FEN GRASS-OF-PARNASSUS

Parnassia glauca

Bittersweet family (Celastraceae)

Description: Perennial; flowering stems are 9"–18" tall. Basal leaves are oval, 1"–2½" long, have round or heart-shaped bases, and are smooth along the margins. Leaves are somewhat stiff and fleshy, and a single leaf clasps the flowering stalk. A single white flower, 1"–1¼" wide, borne at the end of a flowering stalk, has 5 sepals and 5 petals. Petals have light green veins on the inner surface; along with 5 stamens, there are up to 15 or more sterile stamen-like structures (staminodia), which are tipped with nectar-producing glands. Fruit is a ⅓"-long capsule.

Bloom season: Late summer to early fall

Range/habitat: Northeastern United States and northeastern Canada in limestone soils in fens, marshy edges, lake edges, hillside seeps, and springs

Comments: *Parnassia* is from a sixteenth-century name for an unrelated plant growing on Mount Parnassus in Greece, but refers to this plant's high-elevation, moist habitat preference. *Glauca* ("white") refers to the flower color. The sterile stamens are divided at the base into 3 prongs. A variety of flies, bees, and butterflies visit the flowers for nectar or pollen. A female miner bee (*Andrena parnassia*) deposits a "loaf" of nectar and pollen in an underground chamber along with an egg; the developing larvae will feed on this food source.

JIM FOWLER

SESSILE BELLWORT

Uvularia sessilifolia

Autumn Crocus family (Colchicaceae)

Description: Perennial, stems 4"–15" tall; the main stem forks higher up. Oval, strap-shaped leaves are up to 3" long, pointed at both ends, hairless, and stemless. Edges are rolled up at first then flattened with maturity. White to cream flowers, about 1" long, are borne on short flowering stalks either singularly or in pairs and hang downward. The unfused petals curve slightly at the tip. Fruit is a ¾"- to 1¼"-long 3-angled capsule.

Bloom season: Mid-spring to early summer

Range/habitat: Central and eastern North America in moist thickets and woods

Comments: *Uvularia* is from *ūvula* ("little grape"), in reference to the pendulous flowers. *Sessilifolia* ("leaves without petioles") refers to the stemless leaves. Spreading roots shoot up plants that often form dense ground covers or large patches. Early herbalists believed the plants to be beneficial for treating throat ailments because the pendulous flowers resemble an uvula. The young leaves were harvested and cooked for greens. Also known as little merrybells or wild oats. Perfoliate bellwort (*U. perfoliata*) is similar, but the leaves surround the stems (perfoliate).

CHAPPARAL FALSE BINDWEED

Calystegia occidentalis
Morning-Glory family (Convolvulaceae)

Description: Perennial vine that may twist and climb up and over other vegetation. Stems are covered with fine hairs. Arrow- or spade-shaped leaves are about 2" long, covered with fine hairs, lobed at the base, and pointed or rounded at the tip. Basal lobes may be as long as the main lobe. Funnel-shaped white flowers, 1–4, are borne along a single stem. The 5-petaled flowers are 1"–2½" wide. Fruit is a capsule.

Bloom season: Late spring through summer

Range/habitat: Oregon and California in shrub steppe, dry chaparral, and forest edges

Comments: *Calystegia* is from the Greek *kalux* ("calyx") and *stege* ("a covering"), referring to the habit of the petals to close over the capsule. *Occidentalis* ("western") refers to the plant's distribution. Small butterflies, bees, and beetles are common pollinators of these flowers. The flowers are similar to those of *Convolvulus* species, but the false bindweeds have smooth pollen.

CHUCK TAGUE

MOONFLOWER

Ipomoea alba
Morning-Glory family (Convolvulaceae)

Description: Perennial vine, 15'–90' (10'–15') long, with twining stems that have some sharp spines and a milky latex. Leaves are heart-shaped, entire or 3-lobed, 2"–8" long, and arise on a long stem. Funnel-shaped white or pink flowers are 3¼"–6" wide. Fruit is a capsule.

Bloom season: Midsummer to fall

Range/habitat: Tropical areas of North and South America in woodlands

Comments: *Ipomoea* is from the Greek *ips* ("worm") and *homoios* ("resembling") and may refer to either the sprawling roots or twisting nature of the vines. *Alba* ("white") refers to the flower's color. The common name refers to the moonlike shape of the open flowers and because they bloom at night, attracting sphinx moths to pollinate the flowers. In some non-subtropical habitats, this plant may be an annual and only grow 10'–15' tall.

ELEANOR DIETRICH

SOUTHERN DAWNFLOWER

Stylisma humistrata

Morning-Glory family (Convolvulaceae)

Description: Perennial vine. Stems trail along the ground; 1½'–8' long, with fine hairs. Leaves are elliptical to oblong, smooth along the edges, alternate, hairy or smooth, and 1"–2" long. White, funnel-shaped flowers are ¾" long and are borne in leaf axils in groups of 1–3; the styles are fused about one-third their length.

Bloom season: Late spring to fall

Range/habitat: Southeastern United States in waste places, sandy hummocks, floodplains, roadsides, and woodlands

Comments: *Stylisma* is from the Greek *styl* ("pillar" or "column") and *is* ("equal"), referring to the divided style that is equal in length. *Humistrata* is from the Latin *humi* ("earth") and *strat* ("a covering or layer") and refers to the plant's growth habit.

CHRISTIAN GRENIER

CANADIAN BUNCHBERRY

Cornus canadensis

Dogwood family (Cornaceae)

Description: Perennial, low growing, 4"–9" tall. Often spreads by rhizomes. Oval to elliptical leaves are glossy, 1"–2" long, and arranged in small whorls near the end of the stem. A cluster of small greenish-white flowers is surrounded by 4 white, petallike bracts. Fruit resembles a ¼"-wide red berry, but the fleshy coating surrounds a stony cover that houses a single seed, known as a drupe.

Bloom season: Late spring to midsummer

Range/habitat: Across much of North America in coniferous or deciduous woodlands

Comments: *Cornus* is from the Latin name for the Cornelian cherry (*C. mas*). *Canadensis* ("of Canada") refers to the distribution of this plant. Western bunchberry (*C. unalaschkensis*) is similar looking. The fleshy portion of the fruit is edible either raw or cooked; the clusters of red berries give this species its common name.

COASTAL MANROOT

Marah oregana

Gourd family (Cucurbitaceae)

Description: Perennial, with sprawling or climbing stems with branched or coiled tendrils. Irregularly lobed, maplelike leaves are large (up to 10" wide) and have stiff hairs on the upper surface. Bell-shaped male flowers, ½"–1" wide and white, are borne in loose clusters. The fused petals (5–8) form a short tube that flares open at the end, often resembling a 5-pointed star. Male and female flowers are borne separately but on the same plant; female flowers are small, green, and inconspicuous. Fruit is a rounded, fleshy melon that may or may not have weak spines.

Bloom season: Spring to midsummer

Range/habitat: Western North America in open, grassy fields, thickets, bottomlands, and rocky areas

Comments: *Marah* ("bitter") is from a Hebrew word that describes the flavor of the pounded roots. *Oregana* ("of Oregon") refers to the type specimen being collected in the Oregon Territory. Native coastal tribes used the plant to treat kidney problems and skin sores. Also known as western wild-cucumber because of the shape of the fruits. This plant is related to gourds and cucumbers, but the fruit is considered inedible; the bitter leaves are edible.

ELEANOR DIETRICH

GALAX

Galax urceolata

Diapensia family (Diapensiaceae)

Description: Perennial, up to 18" but may be taller; often spreading by rhizomes. Basal leaves are round, shiny, leathery, evergreen, and may turn purplish or reddish brown in winter. The 1"- to 3"-wide leaves have rounded teeth along the margins and a cleft at the bottom, which makes them look heart-shaped. Flowers are borne in a long spike, 8"–15" tall, and are ¼" wide with 4–5 petals. Fruit is a small, urn-shaped capsule with many seeds.

Bloom season: Late spring to midsummer

Range/habitat: Eastern North America; mainly in the Southeast in shady forests and rocky glades

Comments: *Galax* is from the Greek *gala* ("milk") and refers to the flower color. *Urceolata* ("pitcher" or "urn") is in reference to the shape of the capsule. The leaves are often harvested for the floral industry. Also known as beetleweed or wandplant, for the long spike of flowers. Galax, Virginia, gets its name from this plant.

CANDYSTICK

Allotropa virgata
Heather family (Ericaceae)

Description: Mycotroph; grows 5"–18" tall. Stem is red with white stripes (resembles a candy cane) and bears whitish, scalelike leaves. Flowers, borne at the top of the stalk, are urn-shaped, reddish white, and have 5 sepals and 10 stamens. Fruit is a capsule.

Bloom season: Summer

Range/habitat: Western North America in coniferous woodlands

Comments: *Allotropa* is from the Greek *allos* ("other") and *tropos* ("to turn"), in reference to the flowers, which are oriented around the stalk rather than on one side like the related Indian pipe (*Monotropa uniflora*). *Virgata* ("twiggy") refers to the stature of the plant. Look for crab spiders lurking on the flowers, waiting to prey on unsuspecting pollinators. The plants derive nutrients and moisture from other plants through a fungal association; hence its being a mycotroph. Also known as sugarstick or barber's pole, for the white and red stripes.

KINNIKINNICK

Arctostaphylos uva-ursi
Heather family (Ericaceae)

Description: Perennial; low growing up to 10" tall, with stems that mostly trail along the ground. Evergreen leaves are oval to spoon-shaped, leathery, with entire margins. Urn-shaped flowers are pinkish white, ¼" long, and hang downward in clusters. Fruit is a red berry.

Bloom season: Mid-spring to midsummer

Range/habitat: Widespread across the United States, as well as into Guatemala in Central America, in open rocky or sandy areas in clearings or woodlands

Comments: *Arctostaphylos* is from the Greek *arktos* ("bear") and *staphyle* ("bunch of grapes"), which refers to the abundant fruits. *Uva-ursi* ("bear grape") also refers to the cluster of fruits. "Kinnikinnick" is an Algonquin name for an herbal smoking mixture; the leaves were picked for smoking and often added to tobacco. The mealy berries were collected and eaten by Native Americans as well as by wildlife; hence another common name is red bearberry. Alpine bearberry (*A. alpina*) is somewhat similar and grows on New England mountaintops.

JIM FOWLER

STRIPED WINTERGREEN

Chimaphila maculata
Heather family (Ericaceae)

Description: Perennial, 4"–10" tall when in flower; central stem is red or brown. Variegated evergreen leaves are ¾"–2¾" long, oval or lance-shaped, and have distinct white veins. Underside of the leaf is pale, and there are often 2 pairs of basal leaves and a whorl of 3 leaves from which the flowering stalk arises. White to pinkish flowers are borne in small clusters and are ¼"–⅓" wide, with reflexed petals exposing the stamens and pistil below. Fruit is a capsule.

Bloom season: Late spring to early fall

Range/habitat: Eastern North America into Central America in dry, acidic soils in woodlands

Comments: *Chimaphila* is from the Greek *cheima* ("winter weather") and *phileo* ("to love"), in reference to the plant's evergreen leaves, which last throughout the winter. *Maculata* ("spotted") refers to the mottled leaves. Native peoples made a tea from the leaves to treat kidney stones, rheumatism, and stomach ailments. Also called spotted wintergreen or striped prince's pine.

PIPSISSEWA

Chimaphila umbellata
Heather family (Ericaceae)

Description: Perennial, 4"–12" tall, with a woody base and evergreen leaves. Leaves are narrowly rectangular, with sharp teeth along the upper margin, arranged in a whorl pattern, and 2"–4" long; teeth also have small hairs. Saucer-shaped flowers (3–15) hang downward and are a waxy, whitish pink to rose. Flowers are borne in loose clusters; the petals recurve to expose the numerous bilobed stamens and stout pistil. Fruits are round capsules that contain tiny seeds.

Bloom season: Summer

Range/habitat: Widespread in temperate regions of North America in dry coniferous or deciduous forests

Comments: *Chimaphila* is from the Greek *cheima* ("winter weather") and *phileo* ("to love"), in reference to the plant's evergreen leaves, which last throughout the winter. *Umbellata* ("with an umbel") refers to the arrangement of the flower clusters. These plants derive some of their nutrients from association with soil fungi; hence these plants are grouped with other myco-heterotrophs. The astringent leaves of many species in the wintergreen group were used medicinally as a tonic or diuretic. Also known as prince's pine; pipsissewa translates to "flower of the woods," which describes this ground cover.

JIM FOWLER

WINTERGREEN

Gaultheria procumbens
Heather family (Ericaceae)

Description: Perennial, 3"–8" tall; creeping ground cover. Evergreen leaves are elliptical to oblong, ¾"–2" long, and leathery; leaves turn purplish in fall. White, bell- or urn-shaped flowers are ⅜" long, hang downward, and have forked anthers. Flowers may also be tinged with pink. Fruit resembles a red berry but is a capsule enclosed in a fleshy red calyx.

Bloom season: Summer

Range/habitat: Eastern North America in hardwood forests

Comments: *Gaultheria* honors Jean-François Gaultier (1708–1756), a French physician and naturalist who obtained the position of the king's physician in the French colony of Quebec from 1741 to 1756 and was also an avid plant collector and botanist. *Procumbens* ("to lie down") refers to the low-growing growth habit. Also known as boxberry or eastern teaberry, for the edible red fruits, which along with the leaves make a teaberry extract that has been used in flavoring candies, chewing gum, tea, and ice cream.

REUVEN MARTIN

WOOD NYMPH

Moneses uniflora
Heather family (Ericaceae)

Description: Perennial, 4"–6" tall. Oval-shaped leaves are borne in a basal rosette, are ⅓"–1¼" wide, and toothed along the margins. A single nodding white flower, about 2" wide, is borne at the end of an upright stem. Flower has 5 white petals, yellow stamens, and a green style. Fruit is a capsule.

Bloom season: Late spring to early fall

Range/habitat: Much of western and northern North America in damp woodlands and bogs

Comments: *Moneses* is from a Greek word meaning "solitary delight," and *uniflora* means "one flower." Both refer to the beautiful single flower. The fragrant flowers attract bees as pollinators, but the flowers do not produce nectar. Native Americans used the plant to treat colds and rashes.

DAVID LEGROS

INDIAN PIPE

Monotropa uniflora
Heather family (Ericaceae)

Description: Mycotroph. Stems, 4"–10" tall, often arise in clusters and turn black at maturity. The plant lacks green chlorophyll pigments and therefore green leaves or stems. The whitish, overlapping scalelike leaves are narrow to oval-shaped. The single, 1"-long flower is white and bell-shaped and either hangs downward or to the side until mature. When mature, the flower points upward. Fruit is a capsule.

Bloom season: Mid- to late summer

Range/habitat: Western North America in humus-rich sites in dense, moist forests

Comments: Both *Monotropa* ("one direction") and *uniflora* ("one flower") refer to the plant's single flower. The common name indicates the plant's resemblance to a white clay pipe. Since the plant lacks chlorophyll, it derives nutrients from mycorrhizal fungi associations with its roots.

COMMON CRANBERRY

Vaccinium macrocarpon
Heather family (Ericaceae)

Description: Perennial; low-growing or prostrate stems, up to 1' long from a woody base. Leathery leaves are oblong to lance-shaped, change from bronze to green to copper during the season, have blunt round tips, and are ¼"–½" long. Small, white to pinkish flowers are borne in small clusters; individual flowers are ¼"–⅓" long, tube-shaped, have 4 reflexed petals that expose the reddish stamens and single style. Fruit is a dark red berry, ⅓"–½" wide.

Bloom season: Late spring to early summer

Range/habitat: Central and eastern Canada and north-central and northeastern United States in wet bogs, fens, sphagnum marshes, and edges of rivers and lakes

Comments: *Vaccinium* is the classic Latin name for related plants such as bilberry. *Macrocarpon* ("large fruit") refers to the size of the fruits. Small cranberry (*V. oxycoccos*) grows across the northern United States and Canada. Cranberries were first known as "craneberries," for the flower's resemblance to the swept-back wings and long bill of a crane. Also known as American cranberry or bearberry. Commercial varieties are grown for the cranberry industry.

ELEANOR DIETRICH

TEN-ANGLE PIPEWORT

Eriocaulon decangulare
Pipewort family (Eriocaulaceae)

Description: Perennial, 1'–3½' tall. Grasslike leaves, up to 10" long, arise from a basal rosette. Tiny white flowers are arranged in a button or hemispherical-like arrangement atop a leafless stalk that is ribbed and covered with hairs. Heads are ½"–¾" wide. Fruit is an elliptical seed.

Bloom season: Late spring to fall

Range/habitat: Southeastern United States, Mexico, and south to Nicaragua in savannas, bogs, wet prairies, freshwater marshes, woodlands, and ditches

Comments: *Eriocaulon* is from the Greek *erion* ("wool") and *caulos* ("a stalk"), in reference to the dense hairs at the base of the flowering stalk. *Decangulare* ("ten angle") refers to the number of ribs, or angles, that are often present on the flowering stalk. The roots, which are underwater or covered with debris, have air pockets that allow for the absorption of carbon dioxide to aid in photosynthesis. Also known as bog buttons or hatpins, for the hatpin-like resemblance of the flower head and stalk.

TOM LEBSACK

TEXAS BULLNETTLE

Cnidoscolus texana
Spurge family (Euphorbiaceae)

Description: Perennial, 1½'–3' tall, with several stems arising from the same taproot. Stems may be upright or trailing along the ground and exude a milky latex when damaged. Alternate leaves are 2"–4" long, generally 3- to 5-lobed, and covered with stinging hairs. Lobes may be further divided and irregularly toothed along the margins; leaves also have a crinkled appearance. Petalless flowers have 5–7 white sepals, which are fused at their base but spreading like petals. Flowers have 10 stamens and a 3-lobed pistil. Male and female flowers are borne separately but within the same cluster. Fruit is a 3- or 4-chambered, rounded pod covered with stinging hairs.

Bloom season: Spring through summer

Range/habitat: South-central United States and northern Mexico in open fields, disturbed areas, prairies, and woodland edges

Comments: *Cnidoscolus* is from the Greek *knido* (nettle) and *scolus* (thorn), referring to the plant's prickly appearance. *Texana* ("of Texas") refers to the plant's distribution. When contacted, the plant's bristly hairs release an allergic toxin that causes a fierce burning, itching, or stinging pain that may last 30–45 minutes or longer. The seeds are edible, but caution must be taken to harvest the pods due to the stinging hairs. Also known as tread softly (for obvious reasons) or *mala mujer* ("wicked woman") in Spanish.

MATT BERGER

SNOW-ON-THE-MOUNTAIN

Euphorbia marginata
Spurge family (Euphorbiaceae)

Description: Annual, up to 3' tall; often clump forming. When broken, the hairy stems bleed a milky sap. Stalkless leaves are alternate, up to 3" long, and broadly lance-shaped, with a rounded base and pointed tip. Leaves are greenish in spring then may become edged with white, especially those closest to the flower heads. Each flower head bears numerous tiny white to greenish-yellow cup-shaped flowers; each head is about ½" wide. Flowers lack sepals or petals and have white, petallike bracts that subtend the heads. Fruit is a seed.

Bloom season: Summer to early fall

Range/habitat: Much of central and eastern North America in pastures, prairies, and disturbed areas

Comments: *Euphorbia* honors Euphorbus, an ancient Greek physician. *Marginata* ("furnished with a border") refers to the white-lined margins on the showy variegated leaves. The milky sap is toxic and may cause skin reactions or cause poisoning if ingested. William Clark collected the type specimen of this plant while on the Lewis and Clark Expedition of 1803–1806. Snow-on-the-mountain is unique in that the plants may use all three photosynthetic pathways: C3, C4, and CAM (Crassulacean acid metabolism).

STEVE R. TURNER

HOG-PEANUT

Amphicarpaea bracteata
Pea family (Fabaceae)

Description: Annual or perennial, twining vine, 3'–6' long. Compound leaves have 3 leaflets that are diamond- to egg-shaped, pointed at the tip, and with varying degrees of hairs. Middle leaflet is larger than the other 2 leaflets and up to 4" long and 2½" wide. Two types of flowers are borne on stems. White to lilac, ½"-long pea-shaped flowers are arranged in small clusters in the leaf axils. These flowers have 5 lobes; the upper 2 outer lobes curl back over their edges. Smaller, inconspicuous flowers that lack petals are borne on lower, creeping branches and are self-fertilizing. Upper flowers result in inedible seeds arranged in a seedpod that is 1½" long. Lower pods are pear-shaped and may form above- or belowground; these seeds are edible after boiling.

Bloom season: Summer

Range/habitat: Central and eastern North America in moist thickets and woodlands

Comments: *Amphicarpaea* is from the Greek *amphi* ("of both kinds") and *carpos* ("fruit"), referring to the two types of fruit that are formed. *Bracteata* ("bearing bracts") is for the floral stalk's blunt bracts. Hogs dig up and eat the underground fruits.

STEVE R. TURNER

ILLINOIS BUNDLE FLOWER

Desmanthus illinoensis
Pea family (Fabaceae)

Description: Perennial, with 1 to several stems, mainly 1'–4' tall (may reach 5'). The twice pinnately compound, alternate leaves are fernlike and up to 8" long. Each pinnate leaf bears 20–40, ⅛"-long, lance-oblong sub-leaflets with hairy margins. Small white to greenish flowers, 25–50, are borne in lumpy, rounded clusters that are ½" wide and formed at the ends of 3"-long stems. Individual flowers have 5 tiny petals and 5 white stamens that extend past the petals, giving the flower a starburst-like appearance. Fruit is a flat, leathery seedpod about ½" long.

Bloom season: Summer

Range/habitat: Central and eastern United States in prairies, moist meadows, open woodlands, and disturbed ground

Comments: *Desmanthus* is from the Greek *desmos* ("bundle") and *anthos* ("flower"), referring to the flower clusters. *Illinoensis* ("of Illinois") indicates the location of the type specimen of this plant. This odd plant has edible seeds that are high in protein; the root bark is high in the alkaloid methyltryptamine, which is added with beta-carboline compounds to make a hallucinogenic drink called prairiehuasca. The seeds are eaten by game birds and mammals; cattle forage on the leaves. Also known as prairie mimosa because the leaves close up at night or on overcast days, similar to those on a mimosa (*Mimosa* sp.) tree.

WILD LICORICE

Glycyrrhiza lepidota
Pea family (Fabaceae)

Description: Perennial, 1'–4' tall. Compound leaves are 3"–6½" long, with 13–19 lance-shaped leaflets that are pointed at the tips and smooth above. Undersides of the leaves may be slightly hairy or have dotted glandular hairs. Small pea-shaped flowers are arranged along an elongated axis, with 20–50 flowers per axis. Bell-shaped calyx is less than ⅛" long; corolla is white to cream and ⅜"–½" long. Fruit is a seedpod, ⅜"–¾" long and covered with hooked spines.

Bloom season: Mid-spring to late summer

Range/habitat: Across most of North America except the southeastern United States in moist sites along stream banks, fields, meadows, disturbed sites, and roadsides

Comments: *Glycyrrhiza* is from the Greek *glykos* ("sweet") and *rhiza* ("root"), referring to the sweet flavor of the roasted roots, which were eaten by Native Americans. The roots contain the saponin glycyrrhizin. *Lepidota* ("scaly") refers to the brown scales on the leaves. European licorice (*G. glabra*) is commercially used in cough syrups, laxatives, and confections. The hooked seedpods catch on the fur of a passing animal, which helps to disperse the seeds.

STEVE R. TURNER

ROUND-HEAD BUSH FLOWER

Lespedeza capitata
Pea family (Fabaceae)

Description: Perennial, 2'–5' tall. Mostly unbranched; stout stem is covered with hairs. Leaves are trifoliate; ovate to egg-shaped leaflets are up to 3" long and 1" wide, with a prominent midvein. Flower heads arise along an elongated stalk, about 2"–5" long. Each rounded flower cluster (clusters are cloverlike) has numerous hairy bracts that subtend and nearly encase the ½"-long creamy white flowers. The scentless, pea-shaped flowers have 5 petals and a purple spot near the throat. Fruit is a seed.

Bloom season: Midsummer to early fall

Range/habitat: Eastern North America in prairies, savannas, sand dunes, thickets, stream banks, roadsides, wooded areas, and waste places

Comments: *Lespedeza* honors Vicente Manuel de Céspedes (1721?–1794), Spanish Royal Army colonel and Spanish governor of West Florida from 1784 to 1790. He was also known as Vicente Manuel de Zéspedes. *Capitata* ("growing in a dense cluster") refers to the flower heads. Bees are common pollinators of these flowers. Native peoples used the plants to treat rheumatism, as a tea or coffee substitute, and as a poison antidote.

JIM FOWLER

GOAT'S RUE

Tephrosia virginiana
Pea family (Fabaceae)

Description: Perennial, with a semiwoody base and hairy stems, 1'–2½' tall. Feather-like leaves are pinnately compound with 14–28 pairs of oblong or lance-oval shaped leaflets, each about 1" long and covered with fine silky hairs on the undersides. Pea-shaped flowers have a pale yellow to creamy yellow upper petal (the hood) and rose-pink or fuchsia lower outer petals (the wings); the lower central lip (the keel) is a faint yellow-pink. Flowers are about ¾" long. Fruit is a 3"-long seedpod.

Bloom season: Late spring through summer

Range/habitat: Eastern North America in dry savannas, prairies, meadows, and open woodlands

Comments: *Tephrosia* is from the Greek *tephros* ("ash-colored") and refers to the overall silvery appearance of the plant caused by the fine coating of hairs on the leaves and stems. *Virginiana* ("of Virginia") refers to its type locality. The seeds and leaves contain rotenone, and Indigenous peoples ground the plants to create a powder that was sprinkled over the water to kill fish. This forage plant was once thought to increase the production of goat's milk; hence the common name. Also known as catgut or devil's shoestring, for the long, stringy roots.

STEVE R. TURNER

KAREN E. ORSO

DUTCHMAN'S BREECHES

Dicentra cucullaria
Fumitory family (Fumariaceae)

Description: Perennial, 6"–12" tall. Compound, succulent leaves are almost fernlike; stems branch several times, and numerous leaflets also are divided into many segments. Flowers are borne alongside a common stalk; white or pale pink flowers resemble a pair of baggy pants. Outer 2 petals have ½"-long spurs (the pant legs) and flaring tips surrounded by a yellow band (the waistband). Fruit is a seedpod.

Bloom season: Spring

Range/habitat: Mostly eastern North America with disjunct populations in the Pacific Northwest along stream banks, in moist woods, valleys, and ravines

Comments: *Dicentra* ("two spurs") refers to the flower's "baggy pants," which resemble a Dutchman's pantaloons. *Cucullaria* ("hoodlike") also refers to the petals. The sealed flowers are protected from the elements and small pollinators. Large female bumblebees can reach the nectar with their long tongues. Plants contain a hallucinogenic compound; hence another name for these plants is staggerweed, for their effect on livestock. Squirrel corn (*D. canadensis*) is similar but has heart-shaped flowers.

STEER'S HEAD

Dicentra uniflora
Fumitory family (Fumariaceae)

Description: Low-growing perennial, up to 4" tall. Basal leaves are compound and highly divided; there may be 3 to 4 spoon-shaped leaflets. Segments have various lobed tips. The single white to pinkish flower is borne on a short stalk and resembles a steer's head—the 2 outer petals curve outward like horns, and the inner 3 petals create a triangular-looking "skull." Fruit is a seedpod.

Bloom season: Early spring to late summer

Range/habitat: Western North America in meadows, sagebrush flats, and open areas at elevations from 4,900' to 10,000'

Comments: *Dicentra* ("two spurs") refers to the flower's 2 spurred petals. *Uniflora* ("one flower") refers to the solitary flower. These plants are easily overlooked due to their small size. At higher elevations, wildflower enthusiasts should search for the plants (leaves may be more noticeable) along the edges of receding snowlines. Steer's head is a host plant for the larvae of the clodius parnassian butterfly (*Parnassius clodius*).

STEVE R. TURNER

TRACY'S MISTMAIDEN

Romanzoffia tracyi
Waterleaf family (Hydrophyllaceae)

Description: Perennial, arising from a woolly tuber. Leaves are kidney-shaped to round, with broad, scalloped lobes. Basal leaves are 1"–1½" wide with sticky hairs. Flowers are ½" wide and have 5 white petals fused into a funnel-shaped flower. Flowers have a yellow center and barely rise above the leaves. Fruit is a many-seeded capsule.

Bloom season: Summer

Range/habitat: Western North America along coastal bluffs and rocky outcrops

Comments: *Romanzoffia* honors Nikolai Rumyantsev (1724–1826), better known as Count Romanzoff, a Russian sponsor of the Kotzebue expedition to the Pacific Northwest (1815–1920), an expedition to locate a navigable sea route through the Arctic around North America. *Tracyi* is for Samuel Mills Tracy (1847–1920), a botanist with the US Department of Agriculture. Sitka mistmaiden (*R. sitchensis*) is a related species that grows on wet cliffs and rocky outcrops in western mountains.

YELLOW GIANT HYSSOP

Agastache nepetoides
Mint family (Lamiaceae)

Description: Perennial, 4'–7' tall; often with unbranched stems that are 4-angled and generally smooth. Leaves are heart- to egg-shaped (or broadly lance-shaped), opposite, up to 6" long and 3" wide, and have coarse serrations along the margins. Tiny whitish to pale yellow flowers are arranged in dense spikes up to 5" long. The small, 2-lipped flowers have a tubular shape and are about ¼" long. Fruit is a seed.

Bloom season: Midsummer to fall

Range/habitat: Eastern North America and central United States in thickets, disturbed sites, woodland edges, and open woods

Comments: *Agastache* is from the Greek *agan* ("very much") and *stachys* ("an ear of wheat"), in reference to the shape and appearance of the flowering clusters resembling wheat. *Nepetoides* ("like *Nepeta*") refers to this plant's similarity to *Nepeta* species, or catnip. Bees and butterflies are attracted to these plants as pollinators, making them ideal candidates for gardens.

STEVE R. TURNER

NARROW-LEAF MOUNTAIN MINT

Pycnanthemum tenuifolium
Mint family (Lamiaceae)

Description: Perennial, 2'–3' tall on thin, 4-angled stems. Frequent branching creates a bushy effect. Needlelike leaves are opposite, up to 3" long, less than ¼" wide, and hairless. Flowers are arranged in dense, flat-topped clusters at the ends of the stems. Individual, 2-lipped flowers are about ¼" long, with white petals with random purplish spots. Lower lip is 3-lobed. Fruits are small seeds.

Bloom season: Mid- to late summer

Range/habitat: Eastern North America in prairies, moist meadows, thickets, along stream banks, disturbed sites, and open woods

Comments: *Pycnanthemum* is from the Greek *pyknos* ("dense") and *anthos* ("flower"), referring to the dense flower clusters. *Tenuifolium* ("narrow leaves") refers to the narrow leaves. Although the flowers lack an aroma, the foliage has a light minty smell. Some species in the genus were called American wild basil and harvested as herbs.

ELEANOR DIETRICH

AMERICAN GERMANDER

Teucrium canadense
Mint family (Lamiaceae)

Description: Perennial, with downy stems up to 3' tall; numerous smaller side branches may arise in the upper half of the plant. Leaves are opposite; lower leaves are stemmed, upper leaves unstemmed. Leaves are linear or egg-shaped, up to 5" long and 2½" wide, and coarsely toothed along the margins. White or lavender-pink flowers are borne along an elongated stalk that is 1"–5" long. Individual flowers are about ⅓" long; the upper lip is reduced, while the lower lip is enlarged. There are 2 pairs of petallike extensions above the lower lip. Fruit is 4 seeds with fine hairs.

Bloom season: Late spring through summer

Range/habitat: Across much of North America in moist areas such as grasslands, prairies, meadows, thickets, riverbanks, marsh edges, forest edges, and disturbed sites

Comments: *Teucrium* may honor King Teucer of Troy, who first used the leaves as a tea for treating gout and stomach pain. *Canadense* ("of Canada") refers to the plant's distribution. Native Americans harvested the unpleasant-tasting leaves to make an herbal tea to induce sweating or as a diuretic. Modern-day herbalists use the plant for weight control and to lower cholesterol. The common name germander is derived from a Greek word meaning "ground oak," for the leaves of some *Teucrium* species resembling those of oaks. The flower's large lower lip acts like a landing platform for pollinators such as long-tongued bees, flies, and butterflies.

SEGO LILY

Calochortus nuttallii
Lily family (Liliaceae)

Description: Perennial, from an onion-like bulb. Stems are 3"–20" tall, usually with 2–4 linear, grasslike leaves that are 7"–10" long. The striking flowers, which may grow in clusters of 1–3, are 1"–1½" wide, have 3 narrow sepals that are greenish to purplish on the outside and pale inside, and 3 large white, cream, or lavender petals. Petals have pointed tips and a hairy gland located in a patch of yellow on the inside base of the petal. Glands are bordered above by a purple crescent. Fruit is a pointed capsule.

Bloom season: Mid-spring to midsummer

Range/habitat: Western United States in dry grasslands, sagebrush, pinyon-juniper woodlands, and aspen groves

Comments: *Calochortus* ("beautiful grass") refers to the long leaves. *Nuttallii* is for naturalist and Harvard professor Thomas Nuttall (1786–1859), who wrote *The Genera of North American Plants* in 1818 and did extensive plant and animal collections in the western United States. Some years, sego lilies bloom in profusion and cover large patches of desert. This is the state flower of Utah. Native Americans called the plant "sago"; hence the common name of this plant with edible bulbs.

QUEEN'S CUP

Clintonia uniflora
Lily family (Liliaceae)

Description: Perennial. Several broad, elliptical to oblong-shaped leaves, 3"–8" long, arise from an underground stem. Leaves have hairy margins and end in a pointed or rounded tip. A single (rarely 2) white, cup-shaped flower has 6 petallike tepals; flowers are about 1½" wide. Fruit is a blue, bead-like berry.

Bloom season: Late spring to midsummer

Range/habitat: Western North America in moist coniferous woodlands

Comments: *Clintonia* honors DeWitt Clinton (1769–1828), former New York state senator, mayor of New York City, presidential candidate (Peace Party in 1812), and governor of New York, who also wrote natural history books. His political support of the Erie Canal led doubters to label it "Clinton's Ditch." *Uniflora* ("one flower") refers to the single flower borne on the stem. May spread by underground rhizomes. Also known as bluebead lily for the fruit, which forest-dwelling grouse consume. Some Native American tribes mashed the fruit to make a blue dye.

STEVE R. TURNER

WHITE TROUT LILY

Erythronium albidum
Lily family (Liliaceae)

Description: Perennial with short stems, 4"–6" tall. Stems bear 2 broadly lance-shaped leaves that are 3"–6½" long, dark green, and may be mottled with brownish blotches. Immature plants produce 1 leaf and no flowers. A single white, lily-like flower is borne at the tip of the stem and is about 1½" wide and nodding. The 6 tepals curve backward, exposing the white stamens with large yellow anthers. Fruit is a capsule.

Bloom season: Mid- to late spring

Range/habitat: Eastern North America in woodlands

Comments: *Erythronium* is from the Greek *erythros* ("red") and refers to the red dye made from another related plant with pinkish flowers. *Albidum* ("white") refers to the color of the flower. Native Americans and early settlers used the plants for treating gout, ulcers, wounds, nose bleeds, and as a contraceptive. Also known as adder's tongue, dog's-tooth violet (for the shape and size of the corm resembling a dog's tooth), or serpent's tongue (for the shape and patterning of the leaves).

CASCADES LILY

Lilium washingtonianum
Lily family (Liliaceae)

Description: Perennial, 2'–6½' tall. The 1½"- to 5"-long, lance-shaped leaves are arranged alternately on the lower stem and have wavy margins and pointed tips. Upper stem leaves are arranged in whorled patterns along the stout stem. Fragrant, 2"- to 4"-long flowers are bell-shaped, mostly nodding, and have white to pale pink tepals with purplish spots; tepals fade to pink or purple when mature. Fruit is a capsule with numerous seeds.

Bloom season: Late spring to midsummer

Range/habitat: Pacific Northwest in open forests or clearings

Comments: *Lilium* is the Latin name for this genus. *Washingtonianum* honors Martha Washington (1731–1802), wife of George Washington and inaugural First Lady, for whom Albert Kellogg (1813–1887) named this plant in 1859. The very fragrant flowers may be located first by smell; these flowers attract bumblebees and hummingbirds as pollinators. Also known as Washington lily or Shasta lily, these flowers resemble those of the cultivated Easter lily (*L. longiflorum*).

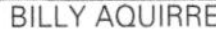

BILLY AQUIRRE

ROCK NETTLE

Eucnide urens

Stickleaf family (Loasaceae)

Description: Perennial, clump-forming, up to 3' tall and 3' wide. Stems, leaves, and buds are covered with sharp, needlelike stinging hairs. Egg-shaped leaves, up to 2½" long, are very hairy, toothed along the margins, and with wrinkled surfaces. White to pale yellow flowers, about 2" wide, have 5 petals that are darker at the base and surround a center mass of about 50 stamens and a white pistil. Mature plants may be cloaked with flowers. Fruit is a small capsule.

Bloom season: Midwinter to early summer

Range/habitat: Southwestern United States in desert washes, hillsides, and rocky outcrops

Comments: *Eucnide* is from the Greek *eu* ("good or true") and *knide* ("stinging nettle"), referring to this plant being strongly like a stinging nettle. *Urens* ("stinging") also refers to the stinging bristles, or hairs, on the plant. Although the stinging hairs deter some herbivores, bighorn sheep eat the leaves.

ELEANOR DIETRICH

SWAMP ROSE MALLOW

Hibiscus moscheutos

Mallow family (Malvaceae)

Description: Perennial with a woody base, grows 3'–7' tall and 2'–4' wide. Broadly egg- or lance-shaped leaves have hairy undersides, toothed margins, and are 3"–8" long; leaves may or may not be lobed. Large, platelike flowers are 6"–9" wide and have 5 white to creamy pink petals with purplish bases that form an inner ring or band. Numerous white to pale yellow anthers surround a long style in the center of the flower. Fruit is a capsule.

Bloom season: Summer to early fall

Range/habitat: Eastern North America in marshes, wetlands, swamps, floodplains, moist woods, and along stream banks

Comments: *Hibiscus* is the old Greek and Latin name for mallow. *Moscheutos* ("musk-scented") refers to the aroma of the crushed leaves and flowers. The seed capsules float on water, which facilitates seed dispersal. Attracted by the nectar or pollen supply, hummingbirds, butterflies, and bees are common pollinators for the flower. This plant is a host for the larval stages of the Io moth (*Automeris io*) and the gray hairstreak butterfly (*Strymon melinus*). Also known as crimson-eyed rose-mallow, for its purplish central ring, or hardy hibiscus, for its stout size.

STEVE R. TURNER

PLAIN'S POPPY-MALLOW
Callirhoe alcaeoides
Mallow family (Malvaceae)

Description: Perennial, 1'–2' tall. Few branched or unbranched stems with star-shaped hairs bear long-stalked lower leaves that are heart- to egg-shaped and lobed or irregularly toothed along the margins. Upper leaves are palmately divided into 5–7 lobes and are 2"–3" long. Cup-shaped flowers are usually borne singly but sometimes in small umbel-like clusters and are pink to white or lavender, 5-lobed, about 2½" wide, and arise from the upper leaf axils. Petals are fringed along the tips. Fruit is a capsule.

Bloom season: Late spring to summer

Range/habitat: South-central and eastern United States in dry rocky or sandy locations in prairies, meadows, roadsides, and disturbed areas

Comments: *Callirhoe* is for the naiad, or nymph, Callirhoe, daughter of Titans Oceanus and Tethys in Greek mythology; one of Jupiter's moons is also called Callirhoe. *Alcaeoides* ("*Alcae*-like") refers to the flowers resembling those in the *Alcae* genus in the Mallow family. Purple poppy-mallow (*C. involucrata*) is similar but has deep purple flowers.

ALEX HEYMAN

ALKALI MALLOW
Malvella leprosa
Mallow family (Malvaceae)

Description: Perennial, with hairy, low-sprawling stems 4"–20" long. Variable-shaped leaves are ⅜"–¾" wide, lobed, hairy, and with wavy or toothed margins; leaves may arise singularly or in groups of 3 from axils along the stem and be triangular, rounded, kidney-, or fan-shaped. Cup-shaped whitish (pale pink to yellowish) flowers are about ¾" long with 5 fused sepals and 5 rounded petals and numerous stamens. Fruit is a disk-shaped structure composed of 6–11 segments, each containing a single seed.

Bloom season: Late spring through summer

Range/habitat: Western United States to Chile in saline or alkaline soils in disturbed sites, edges of vernal pools, or sagebrush steppe

Comments: *Malvella* is a diminutive form of *malva* ("mallow"). *Leprosa* ("scaly") refers to the scalelike hairs on the stems. Butterflies, especially the fiery skipper (*Hylephila phyleus*), are common pollinators attracted to the flowers, although the plants may also reproduce through rhizomes or through broken stems that root.

STEVE R. TURNER

FLY POISON

Amianthium muscitoxicum
Bunchflower family (Melanthiaceae)

Description: Perennial, with leafless flowering stems 12"–36" tall. Grasslike leaves arise from the base and are about 1½" wide and up to 24" long, often arching out and downward. White flowers (which turn bronzy green with age), arranged in dense terminal clusters at the ends of the stems, have 3 petals and 3 sepals. Brownish bracts subtend the flowers. Fruit is a 3-parted capsule.

Bloom season: Early spring to midsummer

Range/habitat: Eastern North America in meadows, fields, savannas, moist ravines, and open woods

Comments: *Amianthium* is from the Greek *amiantos* ("unsoiled") and *anthos* ("flower"), referring to the lack of glands on the base of the petals or sepals. *Muscitoxicum* is from the Latin *muscae* ("flies") and *toxicum* ("poison"), referring to a fly poison made by American colonists by mixing sugar or honey with the crushed bulbs. The plants contain the neurotoxins jervine and amianthine, two toxic alkaloids found throughout the plant, but the bulb is especially poisonous. Beetles are primary pollinators of these plants. Also called crow poison or staggerweed, for the effect on livestock after foraging on the plant.

KATIE BYERLY

MOUNTAIN DEATH CAMAS

Anticlea elegans
Bunchflower family (Melanthiaceae)

Description: Perennial, with grasslike leaves up to 12" long and keeled on the undersides. White flowers are borne in open clusters along an elongated stalk, 2⅓"–5" long. Star- or bowl-like flowers have 6 tepals that have bilobed greenish-yellow glands at their base and notched projections at their tips. Fruit is a capsule.

Bloom season: Summer

Range/habitat: Widespread in North America and northern Mexico except for the southeastern United States in meadows, seeps, hillsides, and montane forests

Comments: In Greek mythology, *Anticlea* is the mother of Odysseus. *Elegans* ("elegant") refers to the plant's stature. The plants are toxic, containing the steroidal alkaloids zygacine and zygadenine, and should not be eaten; hence the common name death camas. Also known as alkali grass or elegant camas. This plant sometimes is included in the *Zigadenus* genus.

EVAN M. RASKIN

FAIRY-WAND

Chamaelirium luteum

Bunchflower family (Melanthiaceae)

Description: Perennial, 2'–4' tall, with both male (2½' tall) and female (4' tall) flowering plants. Both have a basal rosette of 6 (usually) linear, lance- or spoon-shaped leaves that are up to 8" long; upper leaves are smaller. Flowers on the male plants are densely packed; flowering plumes are 4"–9" long and slightly arched. Female flowering plumes are erect and with fewer flowers. Flowers on both plants are very small, less than ⅛" long. Fruit is a capsule.

Bloom season: Late spring to midsummer

Range/habitat: Eastern North America in moist meadows, thickets, and woods

Comments: *Chamaelirium* is from the Greek *chamai* ("dwarf") and *lirion* ("lily"), in reference to the small, lily-like flowers. *Luteum* ("yellow") refers to the yellowish tint of the flower due to the color of the stamens. Also known as devil's bit, squirrel tails, and false unicorn root; the common name fairy-wand is from the wand-like appearance of the flowering stems.

ELEANOR DIETRICH

OSCEOLA'S PLUME

Stenanthium densum

Bunchflower family (Melanthiaceae)

Description: Perennial, up to 30" tall. Narrow, grasslike leaves may be 10" long or longer. White to pink flowers are borne in terminal clusters on elongated stalks rising 3'–5' above the leaves. Small white, saucer-shaped flowers arise on thin stems, are ⅜" wide, and have 5 petals and sepals. Flowering stems and exserted stamens give the flower clusters a feathery appearance. Fruit is a brownish capsule.

Bloom season: Spring to early summer

Range/habitat: Eastern United States in wet flatwoods, moist open savannas, and pine woodlands

Comments: *Stenanthium* ("narrow flower") refers to the narrow sepals and petals. *Densum* ("dense") describes the tightly packed flower clusters. The common name refers to the long plumelike cluster of flowers resembling head feathers worn by Chief Osceola (1804–1838) during the Second Seminole War. Also known as black death camas, black snakeroot, or crow poison, for the toxic leaves and bulbs being used to poison crows feeding on crops.

FEATHER BELLS

Stenanthium gramineum
Bunchflower family (Melanthiaceae)

Description: Perennial, 3'–5' tall. Grasslike leaves arise from a basal rosette, are 7¾"–16" long, and are flat and linear; upper ends of the leaves may be bent. Tiny white flowers are arranged in dense, pyramid-shaped clusters, where the lower flowers are mostly male and the upper ones either female or perfect (both male and female). Flowers are ¼"–½" long, have short stalks, with free sepals and petals. Fruit is an ovoid, 3-lobed capsule.

Bloom season: Summer

Range/habitat: Southeastern United States, from Texas to Pennsylvania, in sandy bogs, wet prairies, along stream banks, or in open woods or forest edges

Comments: *Stenanthium* is from the Greek *stenos* ("narrow") and *anthos* ("flower"), referring to the flower's narrow and pointed sepals and petals. *Gramineum* ("grasslike") refers to the leaves. The common name is derived from the long floral branches, which resemble feathery plumes and bear tiny bell-like flowers. Bees are attracted to the fragrant flowers and are the plant's primary pollinators. Also known as eastern feather bells or grass-leaved lily.

JIM FOWLER

CATESBY'S TRILLIUM

Trillium catesbaei
Bunchflower family (Melanthiaceae)

Description: Perennial, 6"–12" tall. Three oval to broadly lance-shaped leaves are arranged in a whorl around the stem; leaves have deep venation. White to pink 1½"-long flowers arise on a stem that curves under the leaves, and have 3 reflexed petals and protruding yellow stamens. Fruit is a berrylike pod.

Bloom season: Mid-spring to early summer

Range/habitat: Southeastern United States in dry wooded areas

Comments: *Trillium* ("in threes") refers to the number of leaves and floral parts. *Catesbaei* honors Mark Catesby (1679–1749), an English naturalist who studied the flora and fauna of the New World, particularly in the Southeast. He published the *Natural History of Carolina, Florida, and the Bahama Islands* in the 1700s. The seeds have an oil-rich appendage that attracts ants, which help disperse the seeds. Also known as bashful wakerobin.

DAVID LEGROS

LARGE-FLOWERED TRILLIUM

Trillium grandiflorum
Bunchflower family (Melanthiaceae)

Description: Perennial, 1'–1½' tall. From a rhizome arises a single stem, 8"–18", that bears 3 ovate to egg-shaped leaves arranged in a whorl; leaves are 3"–6" long and prominently veined. Arising above the whorl of leaves is a 2"- to 3"-long stalk bearing a single white flower that turns pinkish with age. Flowers are up to 3½" wide and have 3 ovate-shaped petals and 3 smaller green sepals. White petals are reflexed at the tip. Fruit is a berrylike capsule.

Bloom season: Mid-spring to early summer

Range/habitat: Eastern North America in moist thickets or woods

Comments: *Trillium* ("in threes") refers to the number of leaves and floral parts. *Grandiflorum* ("large-leaved") refers to the size of leaves. The seeds have an oil-rich appendage at the tip that attracts ants, which help disperse the seeds. Bumblebees are a common pollinator for these flowers; yellow jackets (*Vespula vulgaris*) and harvestmen (daddy longlegs) also harvest and disperse the seeds. Also known as wood lily or great white trillium; the common names refer to the size, color, and preferred habitat of the flowers.

WESTERN TRILLIUM

Trillium ovatum
Bunchflower family (Melanthiaceae)

Description: Perennial, 8"–20" tall. From a fleshy rhizome arises a single hairless stem that bears large leaves shaped like round-edged triangles in whorls of 3 (may be up to 5). Stalkless leaves are dark green and not mottled. Normally, a single white (turning pink to purplish with age) flower is borne at the end of a short stalk. The 3 flower petals may be up to 2½" long. Fruit is a many-seeded, berrylike capsule.

Bloom season: Spring

Range/habitat: Western North America in moist thickets or in coniferous or deciduous forests

Comments: *Trillium* ("in threes") refers to the number of leaves and floral parts. *Ovatum* ("oval-like") refers to the leaf shape. Ants collect the seeds, which have an oil-rich appendage on the tip that the ants consume. The ants drag the seeds to their nests, and discarded seeds may sprout to form future plants. Trilliums are also called wake-robin for their early springtime appearance, which often coincides with the territorial singing of robins. Meriwether Lewis (1774–1809) collected the type specimen during the Lewis and Clark Expedition, also known as the Corps of Discovery Expedition, along the Columbia River in 1806.

ELEANOR DIETRICH

BEAR GRASS

Xerophyllum tenax
Bunchflower family (Melanthiaceae)

Description: Perennial, 2'–5' tall. Evergreen basal leaves, up to 3' long, grow in clumps and are grasslike, with fine-toothed edges. Stem leaves are similar but get smaller as they progress up the stem. Cream-colored flowers are borne in dense clusters called racemes at the end of the flowering stalk. Individual flowers are borne on long stems, have 6 tepals, and are fragrant. Fruit is a capsule.

Bloom season: Summer

Range/habitat: Western North America in meadows, clearings, burned forests, or undergrowth of dense forests

Comments: *Xerophyllum* is from the Greek *xeros* ("dry") and *phylum* ("leaf"), referring to the non-succulent, persistent leaves. *Tenax* ("tough") refers to the strong leaves, which Native Americans wove into durable capes, baskets, or hats. In spring, bears consume the softer, fleshy leaf bases; hence the common name. The rhizomes of this fire-adapted species can survive a wildfire and then may be one of the first plants to resprout.

SANDBOG DEATH CAMAS

Zigadenus glaberrimus
Bunchflower family (Melanthiaceae)

Description: Perennial, 2'–4' tall, usually unbranched. Grasslike basal leaves are 12"–16" long; upper leaves are smaller. Flowers are borne in clusters of 30–70 white to greenish-white bell-shaped flowers that have 6 tepals and are about 1" wide. Tepals have 2 greenish-gold glands at the base. Fruit is a cone-shaped capsule; flower remains attached as the capsule forms.

Bloom season: Summer to early fall

Range/habitat: Southeastern United States in bogs, savannas, and sandy pine woodlands

Comments: *Zigadenus* is from the Greek *zugon* ("yoke") and *aden* ("gland"), referring to the paired nectar glands on the tepals. *Glaberrimus* ("smooth") refers to the stems and leaves. The common name refers to the type of habitat the plant grows in and the plant's toxicity. There once were about twenty species in the genus; all but this one have been reassigned to other genera.

JIM FOWLER

BUCKBEAN

Menyanthes trifoliata
Buckbean Family (Menyanthaceae)

Description: Perennial, low growing, but from creeping rhizomes that may be very long. Long-stalked, palmately compound leaves bear 3 leaflets up to 4" long on stems 4"–10" long. Leaflets are oval to ovoid-shaped, 1½"–2½" long at flowering time (leaves continue to grow after flowering), and may have wavy margins. Starlike white flowers are borne on long stalks, up to 12" long, and have 5 petals covered with white hairs on the upper surface. Trumpet-shaped flowers are about ½" wide and sometimes have a purplish tint. Fruit is a capsule containing bean-like seeds.

Bloom season: Mid-spring to midsummer

Range/habitat: Temperate North America in shallow waters of bogs, fens, marshes, and ponds

Comments: *Menyanthes* is from the Greek *menyanthos* ("water"), referring to the habitat preference of this emergent aquatic perennial. *Trifoliata* ("with three leaves") refers to the 3 leaflets. In Europe, the leaves were occasionally used as a substitute for hops in beer making or to make schnapps. Also known as bogbean, for the bean-like seeds.

MARGARET MARTIN

LANCELEAF SPRINGBEAUTY

Claytonia lanceolata
Claytonia/Lewisia family (Montiaceae)

Description: Perennial, with plants up to 8" tall; often forming a thick carpet. Basal leaves may or may not be absent; lance- to wedge-shaped stem leaves clasp the upper stem. Leaves are somewhat fleshy. Flower stalks bear 3–20 star-shaped flowers that have white or pinkish petals striped with pink veins and yellow spots near the base. Fruit is a capsule with 2 seeds.

Bloom season: Mid-spring to midsummer

Range/habitat: Western North America in grasslands, shrublands, and forests

Comments: *Claytonia* is for John Clayton (1694–1773), a colonial botanist who collected plants mostly in Virginia and served at one time as the attorney general for Virginia. *Lanceolata* ("lance-shaped") pertains to the shape of the leaves. Plants arise from corms that are edible and said to taste like potato. Miner's lettuce (*C. perfoliata*) is a widespread annual with small white flowers that arise above 2 leaves connected at the base, making them appear as one round leaf pierced by the flowering stalk.

STEVE R. TURNER

VIRGINIA SPRINGBEAUTY

Claytonia virginica

Claytonia/Lewisia family (Montiaceae)

Description: Perennial, with stems 2"–16" tall; often grows in dense clusters. Pairs of grasslike lance-shaped leaves arise along the upper half of the stem; leaves are 1¼"–5½" long; some basal leaves may also be present. White to pink, 5-petaled flowers are arranged in loose clusters; petals bear dark pink veins. Flowers are ¼"–½" wide and have 5 stamens with pink anthers. Fruit is an oval capsule; the small black seeds have fleshy structures on the tip that attract ants as dispersal agents.

Bloom season: Spring

Range/habitat: Eastern North America in ravines, roadsides, lawns, meadows, prairies, and forests

Comments: *Claytonia* is for John Clayton (1694–1773), a colonial botanist who collected plants mostly in Virginia and served at one time as the attorney general for Virginia. *Virginica* ("of Virginia") refers to where the plant was first collected for science. The plants arise from a potato-like corm that is edible and reportedly tastes like chestnuts. Bees and flies are more-common pollinators of these flowers than butterflies. Also known as eastern spring beauty or fairy spud.

BITTERROOT

Lewisia rediviva

Claytonia/Lewisia family (Montiaceae)

Description: Perennial, low growing, arising from a fleshy, carrot-shaped root. Basal leaves are fleshy and rounded and often wither before the flowers mature. Upper stem leaves are very small and bract-like. White to rose flowers are showy, 1"–3" wide, and composed of 12–18 petals. Fruit is an egg-shaped capsule.

Bloom season: Mid-spring to midsummer

Range/habitat: Western North America in gravelly, rocky, or sandy soils in grasslands, sagebrush steppe, and woodlands

Comments: *Lewisia* honors Meriwether Lewis (1774–1809) co-leader of the Corps of Discovery Expedition. *Rediviva* ("brought back to life") refers to the ability of the dried roots to sprout, as one of the samples that Lewis collected and brought back for President Thomas Jefferson sprouted in Jefferson's garden. Northwest tribes collected the prized roots in spring and boiled them before eating. Bitterroot is the state flower of Montana, selected in 1895. The flowers open by midday and attract a variety of pollinators.

CHRISTIAN GRENIER

WESTERN STARFLOWER

Lysimachia latifolia
Myrsine family (Myrsinaceae)

Description: Perennial; low growing (mostly 4"–8" high), but may reach up to 15" tall. Egg-shaped leaves, 1"-4" long with a tapered tip, are arranged in whorls around the stem. Leaves (3–8) occur just below the flowers and have pinnate venation. The ½"-wide white to pink flowers have 4–9 (mostly 6–7) pointed petals and are borne on slender stalks with 1–4 flowers in a cluster. Fruit is a rounded capsule.

Bloom season: Late spring to midsummer

Range/habitat: Western, northern, and eastern North America in shaded woods and along meadow edges and roadsides

Comments: *Lysimachia* is from the Greek *lysis* ("released from") and *mache* ("strife"), referring to Lysimachus, King of Trace in Asia Minor (306–281 BCE), who released himself from stress by waving a sprig of a related plant to calm a charging bull. *Latifolia* ("wide leaves") refers to the shape of the leaves. The common name refers to the star-shaped arrangement of the petals. Indian cucumber-root (*Medeola virginiana*) may be confused with this starflower if the plants are not flowering; however, Indian cucumber-root has parallel venation compared to the pinnate venation of the starflower's leaves.

SEA MILKWORT

Lysimachia maritima
Myrsine family (Myrsinaceae)

Description: Perennial, with prostrate to erect thick stems, 2"–16" long. Fleshy leaves are opposite or alternate, linear to oblong, and ½"–2" long. Flowers lack petals and have a white to pinkish or lavender calyx of 5 petallike sepals and 5 stamens. Fruit is a capsule.

Bloom season: Summer

Range/habitat: Across much of western and northern North America in coastal areas, salt marshes, and alkaline meadows

Comments: *Lysimachia* is from the Greek *lysis* ("released from") and *mache* ("strife"), referring to Lysimachus, King of Trace in Asia Minor (306–281 BCE), who released himself from stress by waving a sprig of a related plant to calm a charging bull. *Maritima* ("maritime") refers to the plant's coastal distribution. Some Native American peoples consumed the cooked roots and brewed a tea from the leaves as a sleep aid. Also known as black saltwort and sea milkweed.

SAND VERBENA

Abronia fragrans
Four O'clock family (Nyctaginaceae)

Description: Perennial, with stems 7"–32" tall that are smooth or covered with glandular hairs. Leaves are opposite; leaf blades are lance- to egg-shaped or linear, ⅜"–3½" long, and covered with fine sticky hairs. White flowers are borne in dense clusters of 25–80 flowers. Flowers have a tube-shaped corolla, ⅜"–1" long, that is lobed and has a wavy flare at the end. Flowers open in late afternoon. Fruit is a wedge-shaped seed.

Bloom season: Early spring to early summer

Range/habitat: Western United States in sandy soils in prairies, plains, open desert, shrublands, and ponderosa pine woodlands

Comments: *Abronia* is from the Greek *abros* ("delicate"), which refers to the flowers. *Fragrans* ("fragrant") refers to the sweet-smelling aroma of the flowers, which bloom at night to attract hawkmoths; other pollinators such as butterflies may pollinate the flowers before they close up during the day. Sand grains adhere to the sticky hairs on the leaves and may aid in dispersal. Also called snowball sand verbena, for the shape and color of the flower heads.

STEVE R. TURNER

CAROLINA FANWORT

Cabomba caroliniana
Water-Lily family (Nymphaeaceae)

Description: Perennial, emergent aquatic, with stems up to 6' long. Bright green, fan-shaped submerged leaves are divided into numerous linear segments, giving the leaves a feathery appearance. Linear to spatula-shaped leaflets are ⅓"–2¼" long. Other, small oval or diamond-shaped leaves are floating leaves. Small white flowers, about ½" wide, barely rise above the water's surface. There are 2 yellow spots at the base of the petals. Fruit is a seed.

Bloom season: Late spring to early fall

Range/habitat: Central and southeastern United States in swamps, sloughs, and floodplains

Comments: *Cabomba* is a Latinized version of its native Guiana name. *Caroliniana* ("of South or North Carolina") refers to the distribution of this plant. This fanwort is commonly used in aquariums and landscaping water features and may be considered invasive outside its native range. A gelatinous mucous covers the aquatic leaves. Also known as fish grass, cabomba, or Carolina water-shield.

FRAGRANT WATER LILY

Nymphaea odorata

Water-Lily family (Nymphaeaceae)

Description: Aquatic perennial. Large pad-like leaves are up to 13" wide and have a cleft near their stems that keeps them from being round. Showy flowers are 1½"–6" wide, with more than 25 white (sometimes pinkish) petals that are pointed at the tip and ¾"–4" long, and 4 sepals. The flowers contain numerous stamens, more than 70. Fruit is a leathery capsule.

Bloom season: Midsummer to fall

Range/habitat: Across much of North America in lowland lakes, ponds, and sloughs.

Comments: *Nymphaea* ("water nymph") refers to the plant's aquatic habitat and nymphlike pure white petals. *Odorata* ("scented") refers to the flowers' sweet smell. Eastern tribes ate the leaves, flower buds, roots, and capsules and used the plant for its medicinal qualities to treat internal disorders. The flowers, which open in the morning and close in the afternoon, contain no nectar; however, beetles are attracted to the flowers for the abundant pollen.

DWARF EVENING PRIMROSE

Oenothera caespitosa

Evening Primrose family (Onagraceae)

Description: Perennial, low growing. Leaves form a basal rosette and are ½"–8" long, long-stemmed; leaf blade is toothed, lobed, entire, or deeply cleft along the margins. White flowers are 2"–3¼" wide with a tube 1"–5" long. The 4 heart-shaped petals are lobed and fade to pink at maturity. Yellow stamens and a 4-pronged style extend far above the flower's throat. Fruit is a woody, rough-textured capsule.

Bloom season: Spring through summer

Range/habitat: Western and central North America in a wide variety of habitats, including open deserts, desert scrub, pinyon-juniper woodlands, and coniferous woodlands at higher elevation

Comments: *Oenothera* is from the Greek *oinos* ("wine") and *thera* ("to seek or imbibe"), referring to the roots of a related species, which were added to flavor wine. *Caespitosa* ("low growing") refers to the stature of this plant. The flowers open in the late afternoon and evening and attract nocturnal insects such as hawkmoths as pollinators. The plant is a larval host for the white-lined sphinx moth (*Hyles lineata*). Also known as tufted evening primrose or gumbo lily.

CRAIG MARTIN

SCARLET BEEBLOSSUM

Oenothera suffrutescens
Evening Primrose family (Onagraceae)

Description: Perennial, 1'–2' tall. Lance-shaped leaves are very narrow, up to 3" long, and with several irregularly spaced teeth along the margins; leaves may also have wavy margins and usually have a covering of fine hairs. Red flowers, up to 12, are borne along an elongated stalk that is up to 16" long. The 4 sepals curve backward toward the stem; the 4 spoon-shaped petals, which point upward, start out white then fade to pink and red as they mature. Pendulous stamens (8) bear reddish-brown anthers; the style is 4-pronged. Fruit is a capsule.

Bloom season: Early spring through summer

Range/habitat: Across much of western North America except Oregon and into northern Mexico in dry prairies, disturbed areas, and woodlands

Comments: *Oenothera* is from the Greek *oinos* ("wine") and *thera* ("to seek or imbibe"), referring to the roots of a related species, which were added to flavor wine. *Suffrutescens* ("slightly woody") refers to the base of the stems. The white flowers attract hummingbird or sphinx moths as pollinators; the flowers fade in color within about 24 hours. Medicinally, the plant has been used as an anti-emetic or childbirth aid.

MARGARET MARTIN

PHANTOM ORCHID

Cephalanthera austiniae
Orchid family (Orchidaceae)

Description: Perennial, with waxy white stems 7½"–24" tall. Lacking leaves, white bracts clasp the stem. White flowers are arranged along an elongated stalk; flowers have yellow markings on the lobed lower lip (the labellum). Dorsal sepal and side petals form a hood over the column, which contains the stamens and styles. Fruit is a capsule.

Bloom season: Late spring through summer

Range/habitat: Western North America in shaded areas along stream banks and woods

Comments: *Cephalanth*era is from the Greek *kephale* ("head") and *anthera* ("anther"), referring to the headlike anthers. *Austiniae* honors Stafford Wallace Austin (1862–1931), a prolific plant collector in California. A myco-heterotrophic plant, the phantom orchid is wholly dependent on its symbiotic relationship with mycorrhizal fungi for its nutrients, as the plant lacks chlorophyll and the ability to photosynthesize. Sweat bees (*Lasioglossum* sp.) are a group of insects attracted to the flowers as pollinators.

STEVE R. TURNER

SPOTTED CORALROOT

Corallorhiza maculata
Orchid family (Orchidaceae)

Description: Perennial, often with numerous upright, yellowish-red or brown stems, 8"–18" tall. Plants lack green leaves but have small translucent scales along the stem. Pink or reddish flowers are borne in a loose cluster at the top of the stem. There are 3 reddish sepals and 3 petals; the lower petal forms a white 3-lobed lip with several reddish spots; flowers may also be yellow-green or purplish. The other 2 petals are reddish and arch around and over the third petal. Fruit is a capsule with numerous seeds.

Bloom season: Late spring and summer

Range/habitat: Widespread across much of North America in humus or forest duff in montane woodlands

Comments: *Corallorhiza* is from the Greek *korallion* ("coral") and *rhiza* ("root"), referring to the knobby, coral-like roots. *Maculata* ("spotted") refers to the crimson spots on the flower's lower lip. The spotted lip offers a wide landing platform for insect pollinators such as flies and small bees. As the insects search for nectar, they are dusted with pollen from the stamens located beneath the flower's upper hood.

JIM FOWLER

RAM'S HEAD LADY-SLIPPER

Cypripedium arietinum
Orchid family (Orchidaceae)

Description: Perennial, 4"–16" tall. Bluish-green leaves (3–5) are narrowly elliptic or lance-oval shaped and 2"–4½" long. Flowers are usually borne singularly or in pairs and are ½"–1" long. The lower lip is conical or like an upside-down pouch, pointed at one end, with red veins, covered along the opening with thin white hairs. The dorsal sepal rises up and forms a hood over the lower petal. The purple or green lateral petals and sepals, which are similar, form spirals that twist away from the center. Fruit is a capsule.

Bloom season: Late spring to early summer

Range/habitat: Northeastern and central Canada into the northern United States in swamps, bogs, and mixed or coniferous forests

Comments: *Cypripedium* is from Greek and Latin words meaning "sandal or shoe of Venus," in reference to the slipper-like appearance of the flowers. *Arietinum* ("small ram") refers to the resemblance of the flower to a ram's head. Although the vanilla-scented flowers attract bees as pollinators, the plant reproduces vegetatively through offshoots of the rhizomes.

CLAIRE WEISER

MOUNTAIN LADY'S SLIPPER

Cypripedium montanum
Orchid family (Orchidaceae)

Description: Perennial, with stems 8"–30" tall. Leaves are broadly elliptic, 2"–6" long and up to 2¾" wide, alternate, and adhering directly to the main stem. Flowers (1–3) are borne near the top of the stem; sepals and petals are maroon-brown, and the lower lip resembles a white pouch.

Bloom season: Late spring to midsummer

Range/habitat: Northwestern United States and western Canada in open slopes, thickets, mixed or coniferous woodlands, and alpine meadows

Comments: *Cypripedium* is from Greek and Latin words meaning "sandal or shoe of Venus," in reference to the slipper-like appearance of the flowers. *Montanum* ("of the mountains") refers to the plant's distribution. The Orchid family is very diverse, with more than an estimated 35,000 species worldwide. Also known as white lady's slipper and moccasin flower, for the color and shape of the flower's pouch.

STEVE R. TURNER

DOWNY RATTLESNAKE PLANTAIN

Goodyera pubescens
Orchid family (Orchidaceae)

Description: Perennial, 4"–18" tall; often spreading by short stolons. Most of the plant, except the petals, is covered with fine hairs. Elliptical to broadly oval-shaped leaves with variegated surfaces form a basal rosette; each leaf is up to 2½" long. Flowering stalk arises from the basal leaves and bears a spike of 10–60 flowers about ¼" long, with white with green or brown markings and a sack-like lower lip.

Bloom season: Midsummer to fall

Range/habitat: Eastern North America in acidic soils in moist to dry forests

Comments: *Goodyera* honors English botanist John Goodyear (1592–1664). *Pubescens* ("downy") refers to the fine hairs that cover most of the plant. The common name refers to the snakeskin-like pattern of the silvery veins and white stripe down the middle of the evergreen leaf's surface. Native Americans used the roots of the plant to treat snakebites and for other medicinal uses. This plant may also be epipetric, growing on rocky shelves.

JAY HORN

MICHAUX'S ORCHID

Habenaria quinqueseta
Orchid family (Orchidaceae)

Description: Perennial, generally 6"–18" tall, but may be up to 2¼' tall. Lance-shaped leaves (3–7) are glossy green, somewhat succulent, 2"–6" long, and alternately arranged. Clusters of 15–25 white flowers are borne loosely along an elongated stalk about 1' long. Each flower is 1½"–2" wide, with white petals and light green sepals that have a dark green stripe. The lower lip is divided into 3 narrow segments with a club-like spur that is 2"–4" long. Fruit is a capsule.

Bloom season: Mid- to late summer

Range/habitat: Southeastern United States from Texas to Florida, the West Indies, and south into South America in marshes, damp woodlands, and open pine woodlands

Comments: *Habenaria* ("rein") is in reference to the floral spur of the type species, which resembles a rein. *Quinqueseta* ("five bristles") refers to the narrow segments of the sepals and petals. This orchid honors André Michaux (1746–1802), a French botanist and explorer who first described this species. Moths are the primary pollinators of these flowers, which have an aroma reminiscent of magnolias; sacs of pollen attach to the moth's compound eyes or proboscis to be transferred to another flower. Also known as false longhorn rein orchid.

DAVID LEGROS

WHITE-FRINGED ORCHID

Platanthera blephariglottis
Orchid family (Orchidaceae)

Description: Perennial, generally 12"–16" tall, but may reach 30". Lance-shaped leaves (2 to several) are glossy green, 3"–8" long, keeled, and alternately arranged along the stem. Flowers are borne in a semi-loose cluster; individual white flowers are ½"–¾" long, with a fringed lower lip and a long curved spur. Lateral sepals bend downward and outward. Fruit is a capsule.

Bloom season: Midsummer to fall

Range/habitat: Eastern North America in bogs, fens, moist prairies, marshes, and pine savannas

Comments: *Platanthera* ("broad anther") refers to the shape of the anthers. *Blephariglottis* is from *blepharis* ("eyelash") and *glottis* ("tongue"), referring to the heavy fringe on the lower, tongue-like lip. Pollinators include bees, butterflies, and moths; sacs of pollen adhere to the insect's compound eyes as it probes the flower for nectar and are transferred to other flowers.

ED ALVERSON

WHITE BOG ORCHID

Platanthera dilatata
Orchid family (Orchidaceae)

Description: Perennial, up to 30" tall; bearing leaves along the entire stem. The 2–5 leaves are oblong to lance-shaped, 3"–8" long, and smaller toward the top of the stem. The small, ¼"- to ½"-wide flowers are white to pale green, very fragrant (smelling like cloves or cinnamon), and arranged in a dense candle-like cluster. The upper sepals and 2 petals form a hood; the lower 2 petal-like sepals spread outward. The lower petal forms a wide lower lip; behind this lip is a slender spur (length varies, depending on the subspecies). Fruit is a many-seeded, egg-shaped capsule.

Bloom season: Summer

Range/habitat: Western and central North America in fens, bogs, swamps, meadows, stream edges, and marshes

Comments: *Platanthera* ("broad anther") refers to the shape of the anthers. *Dilatata* ("spread out") refers to the lower sepals. Native peoples would wash themselves with this sweet-scented flower for good luck. Skippers or noctuid moths with long proboscises such as the owlet moth (*Mesogona olivata*) can reach nectar deep in the flower's spur and are the primary pollinators for these orchids. Similar-looking orchids in the *Habenaria* genus occupy more tropical habitat, while *Platanthera* species grow in more temperate zones. Also known as scent bottle or bog candle, for the upright, candle-like arrangement of the flowers.

ELEGANT PIPERIA

Platanthera elegans
Orchid family (Orchidaceae)

Description: Perennial, with flowering stalk up to 3' tall. Leaves (2–5) are mostly basal, 2"–12" long, up to 3" wide, and oblong or broadly lance-shaped. Top of the stout stem is a dense cluster of small white to greenish-yellow flowers, evenly arranged around the stem. The 3 sepals are white with a green midvein; the 3 petals are similar in size and white or green. A slight spur, ¼"–⅔" long, is located beneath the lower lip and generally about twice the size of the lip. Fruit is a capsule.

Bloom season: Late spring through summer

Range/habitat: Western North America in coastal bluffs, open shrublands, and coniferous forests

Comments: *Platanthera* ("broad anther") refers to the shape of the anthers. *Elegans* ("elegant") refers to the stature of these plants. The flowers have a musky aroma, and the basal leaves often wither before the plants flower. Also known as coast piperia or hillside rein orchid; the common name refers to the strap-shaped spur resembling reins. Some authors place this species in the *Piperia* genus, named for Charles Vancouver Piper (1867–1926), an American botanist and agriculturalist who grew up in the Washington Territory and wrote the first botanical guides to flora in the Northwest.

ELEANOR DIETRICH

HAIRY SHADOW WITCH

Ponthieva racemosa
Orchid family (Orchidaceae)

Description: Perennial, with a basal rosette of dark green leaves that are succulent-like, elliptical to inversely lance-shaped, and 2"–4" long. An upright, hairy reddish-brown or greenish stalk bears 20–35 whitish-green flowers that arise horizontally from the stalk and are ¼"–½" wide. Lower lip is concave and very narrow at the base. Petals and sepals have greenish veins. Fruit is a capsule.

Bloom season: Fall to early winter

Range/habitat: Southeastern United States, West Indies, and south into Mexico and South America in moist areas such as swamps, seeps, hammocks, and shady woodlands

Comments: *Ponthieva* honors Henry de Ponthieu (1731–1808), an English merchant who sent plant and fish specimens from the West Indies to British botanist Sir Joseph Banks in 1778. *Racemosa* ("in a raceme") refers to the arrangement of the flowers, which have a faint citrus-like aroma. Small halictid bees may pollinate the flowers, which may also be self-compatible. The common name is for the flower's resemblance to a flying witch.

ELEANOR DIETRICH

NODDING LADIES' TRESSES

Spiranthes cernua
Orchid family (Orchidaceae)

Description: Perennial. Grass- or strap-like leaves (2–6) are 8"–10" long and arise from the base. White flowers are borne along an elongated stalk, 1'–2' long, in 2–4 spiraling rows. The ½"-wide fragrant flowers curve or nod slightly downward and have a ½"-long lip with crimped margins. The 3 sepals and 3 petals are white; the upper sepal and the 2 lateral petals fuse together to form a hood that curves slightly upward and has 3 small lobes at the tip. Fruit is a green pod; the pod may photosynthesize while developing.

Bloom season: Late summer to fall

Range/habitat: Eastern North America and southern United States in swamps, meadows, bogs, and moist ditches or disturbed sites

Comments: *Spiranthes* is from the Greek *speira* ("coil") and *anthos* ("flower"), in reference to the twisted or braided arrangement of the flowers. *Cernua* ("nodding or bowed") refers to the flowers' downward tilt. The spiraling arrangement of the flowers is due to uneven cell growth. The leaves often wither prior to the plant's flowering. Long-lipped ladies'-tresses (*S. longilabris*) has a long lower lip and is one of more than a dozen ladies'-tresses that grow in eastern North America.

KAREN E. ORSO

HOODED LADIES' TRESSES

Spiranthes romanozoffiana
Orchid family (Orchidaceae)

Description: Perennial. From fleshy roots arise 2–5 basal leaves that are strap-like and up to 10" long. Flowering stalk, averaging 2"–6" tall, bears numerous creamy to greenish-white flowers arranged in distinct longitudinal rows. Individual flowers have sticky, hairy sepals and petals that form a hood, with the lower petal bent sharply downward. Fruit is a capsule.

Bloom season: Mid- to late summer

Range/habitat: Across much of western and northern North America in bogs, fens, moist meadows, along stream banks, and in woodlands

Comments: *Spiranthes* is from the Greek *speira* ("coil") and *anthos* ("flower"), in reference to the twisted or braided arrangement of the flowers. *Romanozoffiana* honors Count Romanzoff (Nikolai Rumyantsev, 1754–1826), a Russian count who financially supported Otto von Kotzebue (1778–1846), a Russian officer and navigator in the Imperial Russian Navy, on an expedition to explore Alaska in 1815–1818. This orchid was collected for science during the expedition and named for Count Romanzoff. In Ireland this plant is known as Irish ladies' tresses.

JIM FOWLER

THREE-BIRDS ORCHID

Triphora trianthophoros
Orchid family (Orchidaceae)

Description: Perennial, semi-saprophyte, with stems 1½"–6½" tall. Small, dark green egg-shaped leaves are up to ½" long and clasp the stem. White to pink flowers are ¾" long and in small clusters of 1–8 (mostly 3) flowers. Lower lip is 3-lobed and has green ridges running down the center; the flower lacks a floral spur. Fruit is a capsule.

Bloom season: Midsummer to early fall

Range/habitat: Central and eastern United States in forest litter of thickets, occasionally swamps, and mixed deciduous woods often composed of American beech (*Fagus grandifolia*)

Comments: *Triphora* ("threefold") refers to the 3 flowers and 3 ridges on the flower's lip. *Trianthophoros* ("bearing three flowers") and the common name refer to the number of flowers, often in 3s, and these flowers resembling birds in flight. The short-lived flowers tend to bloom en masse to increase pollination success, and the timing of the blooms is sometimes regarded in folklore as "the week in August following the first drenching rain." Short-lived, the flowers may last just 1 day. Many species of *Triphora* are also known as noddingheads.

SICKLETOP LOUSEWORT

Pedicularis racemosa
Broomrape family (Orobanchaceae)

Description: Perennial, up to 2' tall; stems are reddish when they first appear and resemble a bouquet. Upper lance-shaped leaves are larger than lower leaves and have short stalks. Leaf edges are double saw-toothed. Flowers are whitish but mostly pink to purplish. Flower's upper lip tapers into a sickle-shaped, curved beak; petals in the larger, lower lip spread outward. Fruit is a flat, curved capsule.

Bloom season: Summer

Range/habitat: Western North America in coniferous woods

Comments: *Pedicularis* ("pertaining to lice") refers to the belief that livestock became infected with lice when grazing in fields where members of this genus grew. *Racemosa* ("cluster") refers to the type of floral arrangement in which the stalked flowers are attached along an unbranched stem. The sickle-shaped flowers lend themselves to the common name as well as another name for the plant: parrot's beak.

MATT BERGER

WHITE BEAR-POPPY

Arctomecon merriamii
Poppy family (Papaveraceae)

Description: Perennial, with waxy stems, 9"–20" tall. Basal leaves are very hairy, wedge-shaped, 1"–3" long, and with rounded teeth along the margins. One white, 6-petaled flower is borne singularly on a flowering stem, nodding at first then becoming erect with maturity. Flowers have 3 green sepals covered with shaggy white hairs, numerous stamens, and are 2"–3" wide. Fruit is a capsule.

Bloom season: Mid-spring to early summer

Range/habitat: Southwestern United States in gypsum-rich desert soils

Comments: *Arctomecon* is from *arktos* ("bear") and *mecon* ("poppy"), referring to the hairy leaves resembling a bear's paw and the poppy-like capsules. *Merriamii* honors C. Hart Merriam (1855–1942), an American zoologist and ornithologist who served as the naturalist with the 1872 Hayden Geologic Survey at the age of 16. Also known as desert bearpoppy and great bearclaw poppy.

MATT BERGER

MOHAVE PRICKLY-POPPY

Argemone corymbosa

Poppy family (Papaveraceae)

Description: Perennial, moderately branched and 8"–36" tall, with stout, spiny stems. Stems bleed an orange latex when crushed. Leaves are inversely lance-shaped, 1"–6" long, armed with stout spines below (sparingly above), and lobed halfway to the middle of the leaf. White flowers, 1½"–3½" wide, have numerous (100–120) yellow stamens and crinkled petals. Outer petals are as broad as they are long; inner ones are much broader than long. Capsule is football-shaped and spiny.

Bloom season: Mid-spring to early summer

Range/habitat: Southwestern United States in disturbed areas, sandy sites, dry slopes, and desert shrub

Comments: *Argemone* is from the Greek *argema* ("cataract"), a disorder of the eye this plant was used to treat. *Corymbosa* ("clustered") refers to the flat-topped arrangement of the flowers. This plant was also used as a purgative in substitution for syrup of ipecac. A variety of bees are pollinators for these plants.

SCOTT LOARIE

PRICKLY POPPY

Argemone polyanthemos

Poppy family (Papaveraceae)

Description: Annual, stems 1'–4' tall, with yellow-orange sap and light brown spines. Blue-green leaves are up to 8" long and lobed to nearly the midvein. Undersides of the leaves and margins have few prickles; the upper surface lacks spines. White flowers, up to 5" wide, have numerous stamens (about 150), 3 sepals, and 4–6 wrinkled petals. Fruit is a 2"-long capsule with widely spaced brown spines.

Bloom season: Mid-spring to fall; may bloom in late winter, depending on conditions

Range/habitat: Southwestern and central United States in prairies, grasslands, meadows, plains, disturbed sites, and savannas

Comments: *Argemone* is from the Greek *argema* ("cataract"), a disorder of the eye this plant was used to treat. *Polyanthemos* ("many flowers") refers to the abundance of flowers that may be present. All parts of this plant are poisonous, and, along with the spiny nature of the leaves and stem, it is generally avoided by livestock. This plant has a deep taproot and may be biennial in parts of its range. Native Americans used the sap to treat warts and as a dye to paint their arrow shafts. Also known as crested prickly poppy or fried egg flower.

ALEX HEYMAN

CREAM-CUPS

Platystemon californicus
Poppy family (Papaveraceae)

Description: Annual; highly variable, depending on location; sap may be colorless to orange. Stems are 1¼"–12" long and may be slightly or strongly hairy. Wide, linear- to lance-shaped leaves are whorled or oppositely arranged, ⅓"–3½" long, with rounded or pointed tips. A single flower is borne on a long stem and has 3 hairy sepals and at least 6 petals (larger flowers may have more). Bowl-shaped flowers are about 1" wide, with all white or yellow petals, or white petals with either a yellow tip or base or both; petals fade to reddish as they age. Numerous stamens, some petallike, are arranged in several whorls inside the flower. Fruit is a capsule that resembles a shucked tiny ear of corn.

Bloom season: Spring

Range/habitat: Western United States and Baja California in grasslands and sandy sites

Comments: *Platystemon* is from the Greek *platos* ("broad") and *stemon* ("stamen"), referring to the flattened stamen filaments. *Californicus* ("of California") describes the distribution of this plant. The flowers are pollinated by bees or the wind. The common name refers to the similarity of these flowers to those in the buttercup genus, *Ranunculus*.

KATIE BYERLY

BLOODROOT

Sanguinaria canadensis
Poppy family (Papaveraceae)

Description: Perennial, 6"–9" tall. Roots and stems bleed an acrid yellow-orange juice. A single, palmately lobed basal leaf is generally round in outline, has 5–7 lobes, and is 12"–14" long at maturity. The leaf wraps around the single flower bud, unfurling as the flower blooms. The white, 2"-wide flower has 8–10 petals and numerous stamens forming a yellowish center and arises on a separate stalk. Fruit is a 1½"- to 2¼"-long pod-bearing seed with small fleshy structures (elaiosomes) that attracts ants.

Bloom season: Spring

Range/habitat: Central and eastern North America along streams and in woods

Comments: *Sanguinaria* ("bleeding") refers to the color of the sap when the roots, stems, or leaves are crushed; hence the common name. *Canadensis* ("of Canada") refers to the type locality and distribution of this plant. Some Native American peoples used the reddish juice of the plant as a dye for basketry, clothing, and war paint, as well as an insect repellent. Small flies and bees pollinate the flowers. Also known as redroot or red puccoon.

CHUCK TAGUE

POKEWEED

Phytolacca americana
Pokeweed family (Phytolaccaceae)

Description: Perennial; may grow 4'–10' tall, with green or purple stems. Roots are up to 4" thick and 1' long. Alternate leaves are 5"–15" long, 2"–7" wide, and give off a strong aroma when crushed. Flowers are borne in 8"-long pendulous clusters; individual flowers are ¾" wide, with 5 white or greenish petallike sepals, no petals, and 10 stamens. Fruits are purple or black berries, about ¼" wide, arranged like a cluster of grapes.

Bloom season: Early summer to fall

Range/habitat: Eastern North America and the central United States in moist soils in ditches, thickets, pastures, fields, and open woods

Comments: *Phytolacca* is from *phyton* ("plant") and *lacca* ("crimson"), referring to the color of the fruits and the dye produced by the berries. *Americana* ("of America") refers to the distribution of the plant. The leaves and shoots are edible when young (with proper cooking) but become toxic with age. The berries were used by colonists to enhance cheap wine. In some sites the plant stems may reach up to 21' long.

STEVE R. TURNER

WHITE TURTLEHEAD

Chelone glabra
Plantain family (Plantaginaceae)

Description: Perennial, 2'–4' tall; often clump forming. Lance-shaped leaves are oppositely arranged, either sessile or borne on very short stems, up to 8" long, and toothed along the margins; successive pairs of leaves are arranged 90 degrees from the next pair. White, 2-lipped tubular flowers are borne in tight clusters along an elongated stalk that is 3"–6" long; flowers may have a tinge of lavender or purple and are about 1¼" long. Corolla is somewhat flattened at the mouth, resembling a turtle's head. Fruit is a capsule.

Bloom season: Midsummer to mid-fall

Range/habitat: Eastern North America in swampy areas, along stream banks, and in moist woods

Comments: *Chelone* is from the Greek for "tortoise," referring to the flowers resembling a turtle's head when viewed straight on. *Glabra* ("without hairs") refers to the smooth stems and leaves. This plant is a larval host for the Baltimore checkerspot butterfly (*Euphydryas phaeton*), Maryland's state insect. Pink turtlehead (*C. obliqua*) has pinkish flowers and broader leaves.

JUDY GALLAGHER

WHITE-FLOWERED MECARDONIA

Mecardonia acuminata

Plantain family (Plantaginaceae)

Description: Perennial, up to 6" tall, often with horizontal stems. Elliptic to lance-shaped leaves are toothed or smooth along the margins, have a narrow base, and are oppositely arranged. Tubular flowers arise from leaf axils and are 2-lipped, with 5 white or pink-tinged lobes and a hairy throat with pinkish veins. Fruit is an ovoid capsule.

Bloom season: Summer

Range/habitat: Southeastern United States to South America in bogs, swamps, marshes, savannas, stream banks, and disturbed places

Comments: *Mecardonia* honors Antonio Meca y Cardona (1726–1788), a Spanish military officer and founder of the botanical gardens at Barcelona's Royal College of Surgery. *Acuminata* ("pointed at the ends") describes either the leaf base or the sepal tip. The flowers attract bees as pollinators. Also known as axilflower, for the flowers arising from the leaf axils, or black hedge hyssop, for the plants turning black at maturity.

MARGARET MARTIN

WHITE BEARDTONGUE

Penstemon albidus

Plantain family (Plantaginaceae)

Description: Perennial, with upright stems 8"–12" tall. Stems may be slightly hairy below and with sticky glands above near the flowers. Leaves are mostly basal, oblong to spatula-shaped, hairy or hairless, and 1"–4" long. White, pinkish, or bluish tubular flowers are arranged along an elongated stalk, are ¾" long, and are 2-lipped; inside of the flower has reddish or purplish nectar lines. Fruit is a capsule.

Bloom season: Mid-spring to midsummer

Range/habitat: Central North America in dry prairies or hillsides

Comments: *Penstemon* is from *pen* ("almost") and *stemon* ("stamen"), referring to the sterile stamen (staminode) typical of this genus. *Albidus* ("white") refers to the color of the flowers. The common name beardtongue refers to the one sterile stamen that bears numerous hairs.

STEVE R. TURNER

FOXGLOVE PENSTEMON

Penstemon digitalis
Plantain family (Plantaginaceae)

Description: Perennial, 3'–5' tall; several flowering stems arise from the base. Basal leaves are variable, from elliptical to broadly lance-shaped, and up to 6" long. Stem leaves are lance-shaped, oppositely arranged, stalkless, up to 5" long, and bear small teeth along the margins. Tubular flowers are white to purple-tinged, 1½"–2" long, and are 2-lipped with 5 lobes. Corolla is slightly inflated; flowers are borne in clusters at the top of the stems. Fruit is a capsule.

Bloom season: Mid-spring to late summer

Range/habitat: Eastern North America and southeastern United States in prairies, fields, thickets, open woods, and disturbed sites

Comments: *Penstemon* is from *pen* ("almost") and *stemon* ("stamen"), referring to the sterile stamen (staminode) typical of this genus. *Digitalis* ("fingerlike") refers to the flowers resembling the fingers of a glove. The plants are hairless except for fine hairs on the outer surface of the flowers. Long-tongued bees of several types are common pollinators of these flowers; comblike hairs on the anthers deposit pollen on the bee's back as it brushes against the stamen. Some authorities believe the plant was originally native to the Mississippi Basin but spread with human settlement.

PALMER'S PENSTEMON

Penstemon palmeri
Plantain family (Plantaginaceae)

Description: Perennial, with clustered stems, 2'–6' tall. Thick leaves, ¾"–4½" long, are elliptical and smooth, oppositely arranged along the stem, and have toothed margins. Flowers are creamy white to lavender, 1"–1½" long, inflated, with prominent wine-red nectar guidelines on the inside of the lower lip and a prominent yellow "beard." The 2-lipped flowers have a sterile stamen that protrudes beyond the opening. Fruit is a capsule.

Bloom season: Late spring through summer

Range/habitat: Western United States in disturbed sites, roadsides, canyon bottoms, shrublands, and woodlands

Comments: *Penstemon* is from *pen* ("almost") and *stemon* ("stamen"), referring to the sterile stamen (staminode) typical of this genus. *Palmeri* honors Edward Palmer (1831–1911), an English immigrant who collected many native American plant and bird specimens in the West. Palmer's penstemon is often included in commercial seed mixes and used in landscaping and roadside beautification projects, thus expanding its distribution. This plant is one of the few penstemons with fragrant flowers.

KATIE BYERLY

CULVER'S ROOT

Veronicastrum virginianum
Plantain family (Plantaginaceae)

Description: Perennial, 2'–7' tall. Lance-shaped leaves are arranged in a whorl of 3–7. Each leaf is toothed along the margins, up to 6" long, and about 1½" wide. Flowering stems reach upward to about 9" long and bear numerous small, white to bluish tube-shaped flowers with exserted stamens. As various flowering stems arise from side branches, the overall appearance is similar to a candelabra. Fruit is a seed.

Bloom season: Early summer to early fall

Range/habitat: Northeastern North America in meadows, glades, wet prairies, thickets, and woodlands

Comments: *Veronicastrum* is a combination that refers to these flowers resembling those of *Veronica. Virginianum* ("of Virginia") refers to the plant's distribution and type locality. Although the fresh root is toxic, Native Americans made a tea from dried roots to treat various ailments. The common name refers to Dr. Culver, an eighteenth-century physician who promoted the health benefits of the roots as a purgative and laxative. Also known as black root.

MAREK BOROWIEC

EVENING-SNOW

Linanthus dichotomus
Phlox family (Polemoniaceae)

Description: Annual; thin, waxy stems are about 8" tall and greenish purple. Opposite leaves are divided into linear lobes, each ⅓"–1⅓" long. Night-blooming flowers (vespertine) have 5 white petals with a light purplish underside; petals unfurl to form a funnel-shaped flower. Stamens and style do not project beyond the mouth of the flower. Flowers are often borne singly or in small groups. Fruit is a capsule.

Bloom season: Spring

Range/habitat: Western United States in serpentine or sandy soils along hillsides or in open desert areas

Comments: *Linanthus* is from *linum* ("flax") and *anthos* ("flower"), referring to the flowers resembling those in the *Linum* (flax) genus. *Dichotomus* ("divided or branched into pairs") refers to the leaves. The flowers produce a variety of floral scents across the plant's distribution, attracting night-flying pollinators such as moths.

GRANITE PRICKLY PHLOX

Linanthus pungens
Phlox family (Polemoniaceae)

Description: Perennial; mat-like growth up to 20" long, but may also grow erect. The ¼"- to ½"-long, needlelike leaves form dense clusters along the stems. Leaves are divided into 3–7 segments, similar to fingers on a hand; leaves are opposite below and alternate above. Flowers are funnel-shaped and white and have 5 flaring petals that may be tinged with pink. In bud, the 1"-long flowers are twisted shut. Fruit is a tiny seed.

Bloom season: Spring to midsummer

Range/habitat: Western North America in rocky or sandy sites in mixed shrublands or woodlands

Comments: *Linanthus* is from *linum* ("flax") and *anthos* ("flower"), referring to the flowers resembling those in the *Linum* (flax) genus. *Pungens* ("sharp-pointed") describes the tips of the leaves. The flowers unfurl at night and attract moths as pollinators. Also called prickly phlox or granite gilia.

HOOD'S PHLOX

Phlox hoodii
Phlox family (Polemoniaceae)

Description: Perennial; mat- or cushion-like, mainly 2"–15" wide. Stems and leaves are covered with white hairs. Linear leaves are opposite, tightly arranged, and ⅛"–¾" long, with spiny tips. White, blue, or lavender flowers arise from a hairy calyx. Corolla tube is about ½" long and flares open at the top to form 5 lobes. Fruit is a seed.

Bloom season: Early spring to early summer

Range/habitat: Western North America in sagebrush, desert shrub, pinyon-juniper woodlands, or ponderosa pine forests

Comments: *Phlox* ("flame") is in reference to the brightly colored flowers of many *Phlox* species. *Hoodii* honors Robert Hood (1797–1821), map maker and artist on the Franklin Arctic Expedition of 1819–1822, who was killed by one of the sailors. Hood was one of the first people to accurately describe the aurora borealis as an electrical phenomenon. Desert phlox (*P. austromontana*) is similar but with larger flowers and more lightly colored leaves. Also known as spiny phlox or carpet phlox, for its spiny leaves and mat-like growth.

LEAH BREITENSTINE

WHITE MILKWORT

Polygala alba
Milkwort family (Polygalaceae)

Description: Annual, 4"–16" tall, with numerous upright and hairless stems. The ¼"- to 1"-long lance-shaped or linear leaves are arranged in whorls of 3–7 or alternately along the stem. Flowers are borne along a dense spike that is somewhat conical and about 1"–3" long. Flowers are white with a greenish center, about ⅛" wide, and have 5 sepals, of which 2 are larger and petallike. Of the 3 petals, 2 are united to form a tube that surrounds the 8 stamens and pistil; the other petal has a keel with 8 narrow lobes. Buds are purplish, but the flowers are white. Fruit is an oval-shaped seedpod.

Bloom season: Late spring to early summer

Range/habitat: Central North America in dry prairies, fields, hillsides, open woods, and disturbed sites

Comments: *Polygala* is from the Greek *poly* ("much") and *gala* ("milk"), from the belief that livestock foraging on these plants produced more milk. *Alba* ("white") refers to the color of the flowers. Baldwin's milkwort (*P. balduinii*) has small, dense clusters of white flowers and honors William Baldwin (1779–1819), an American physician and botanist.

JIM FOWLER

SENECA-MILKWORT

Polygala senega
Milkwort family (Polygalaceae)

Description: Perennial; multiple unbranched stems rise up to 12"–20" tall. Alternate leaves are lance-shaped, slightly toothed or hairy along the margins, and up to 3½" long; lower leaves are smaller and scalelike. Flowering stalk bears a thick cluster of white to green tubular-shaped flowers, about ¼" wide. Flowers have 3 petals (1 is fringed), 2 white petallike lateral sepals, 3 light green to purplish non-petallike sepals, and 8 stamens; petals are fused together to form a tubelike corolla. Fruit is a capsule with 2 black seeds.

Bloom season: Late spring to midsummer

Range/habitat: Central and eastern North America in moist or dry prairies, riverbanks, disturbed sites, and open woodlands; often in calcareous soils

Comments: *Polygala* is from the Greek *poly* ("much") and *gala* ("milk"), from the belief that livestock foraging on these plants produced more milk. *Senega* is an alteration of Seneca and honors the Seneca tribe, which historically used this plant to treat snakebites and other ailments. Other northern tribes used the plant medicinally to treat a variety of ailments, including colds, as a diuretic, to soothe toothaches or sore throats, as an expectorant, and for croup. Settlers sent the roots back to Europe, where they were sold to treat pneumonia; the roots are still used medicinally today. Various sizes of bees are the primary pollinators of the flowers.

AMERICAN BISTORT

Polygonum bistortoides
Buckwheat family (Polygonaceae)

Description: Perennial, 10"–35" tall; 1 to several unbranched stems arise from a central base. Stems bear a few long-stalked, lance-shaped or elliptical leaves along the lower portion of the stem. Leaves are 7"–10" long; upper leaves are smaller and stalkless. Small, white to pinkish flowers are borne in dense clusters, 1"–3" long. Fruits are small, 3-angled, yellowish-brown seeds.

Bloom season: Summer

Range/habitat: Western North America in wet meadows or along stream banks

Comments: *Polygonum* ("many knees") refers to the many joints along the stems. *Bistortoides* ("twice-twisted") refers to the plant's thick, twisted root and earns the plant another common name: snakeroot. Native Americans roasted or boiled the roots for food; the flavor is chestnut-like. The plant was first collected for science in 1806 during the Lewis and Clark Expedition. The flowers, which have an unpleasant odor, give rise to another common name: miner's toes. Alpine bistort (*Bistorta viviparum*) is similar but smaller and has bulblets below the flower clusters.

PUSSYPAWS

Calyptridum umbellatum
Purslane family (Portulaceae)

Description: Perennial at higher elevations (sometimes annual at lower elevations), with 2"- to 10"-long flowering stems radiating outward and sprawling over the ground. Club- to spatula-shaped leaves, ½"–2" long, are arranged in a basal pattern and are smooth and shiny. Reddish flowering stalks extend from the leaves and bear a round cluster of fuzzy white to pink flowers. Each ¼"-long flower has 2 round and papery sepals that surround the 4 smaller petals and stamens. Petals wither before the sepals fall off. Fruit is a flattened capsule.

Bloom season: Late spring through summer

Range/habitat: Western North America in dry, sandy or pumice soils in subalpine or alpine environments; also in ponderosa pine forests

Comments: *Calyptridum* ("having a calyptra") is for the cap-like covering of the flower. *Umbellatum* ("flowers in an umbel") describes the arrangement of the flowers in an umbrellalike pattern. During the day the stems may elevate, which protects the flowers from overheating and perhaps makes them more attractive to pollinators. The common name is for the clusters of fuzzy flowers that resemble a cat's upturned paw.

STEVE R. TURNER

EASTERN SHOOTING STAR

Primula meadia

Primrose family (Primulaceae)

Description: Perennial, up to 2' tall. Broadly lance- to inversely lance-shaped basal leaves are up to 12" long and 3" wide, arise on short petioles, and have rounded tips. Clusters of flowers arise on arching flowering stalks that arise from leafless stems, which are green or purple. Downward-pointing shooting star–like flowers are white to dark pink, with 5 reflexed petals, and about 1" long. Petals join together at the base into a tube, with wavy rings of pink-purple, white, yellow, and maroon at the base; petals appear somewhat twisted. Exserted stamens form a beak-like cone and are tightly appressed to the style; the arching flowers become erect post-pollination. Fruit is a thick-walled, dark reddish-brown capsule.

Bloom season: Mid-spring to early summer

Range/habitat: Eastern and central North America in glades, meadows, prairies, rocky ledges, and wooded slopes.

Comments: *Primula* is from the Latin *primus* ("first"), indicating that some members of this genus bloom early in the spring. *Meadia* honors Richard Mead (1673–1754), an English physician who wrote a historically important text on understanding transmission of diseases. Bees are the primary pollinators of these flowers. Some use "buzz pollination," which involves hovering just below the stamens and rapidly vibrating their thoracic muscles to cause pollen to be released. Jeweled shooting star (*Dodecatheon amethystium*) is similar but has thin-walled, yellowish to reddish capsules.

JIM FOWLER

BLACK COHOSH

Actaea racemosa

Buttercup family (Ranunculaceae)

Description: Perennial, 4'–6' tall, but may reach 8' high. Bipinnately or tripinnately compound leaves are deeply dissected, divided into clusters of 3–5 leaflets that are a dark glossy green. Individual leaflets (10–20 per compound leaf) are lance- to egg-shaped, up to 4" long, and toothed along the margins; terminal leaflet is variously cleft. Tiny white flowers are arranged in terminal wispy spikes, 1'–3' long. Flowers are about ⅔" wide, lack petals, have over 20 stamens, and have tiny sepals that drop off early. Fruit is a capsule borne singularly, not in clusters.

Bloom season: Summer

Range/habitat: Eastern North America in shady sites in rocky woods

Comments: *Actaea* is adapted from the ancient Latin name for the plant. *Racemosa* ("in a raceme") defines the flower arrangement. The common name cohosh is from an Algonquin word for the rough-textured rhizomes, which are still collected for their medicinal properties in treating menopausal or premenstrual conditions. Also known as black bugbane, for the plant's insect repellent properties, or fairy candles, for the upright, candle-like clusters of white flowers.

STEVE R. TURNER

RED BANEBERRY

Actaea rubra
Buttercup family (Ranunculaceae)

Description: Perennial, growing to 3' tall. Plants bear few leaves, generally 1 near the base and others higher on the stem. Leaves are divided 2 or 3 times into divisions bearing 3 leaflets. Leaflets are sharply toothed and lobed. Terminal rounded cluster of tiny, white flowers with protruding stamens resembles an elongated white ball. Petallike sepals remain after flowering, while the petals fall away. Exserted stamens give the flower a feathery appearance. Fruits are smooth berries that are green at first, becoming reddish with age; some plants may bear white berries.

Bloom season: Mid-spring to early summer

Range/habitat: Across much of western and northern North America in moist, shady woods

Comments: Baneberry is from the Anglo-Saxon *bana* ("murderous") and refers to the highly toxic compounds in the plant's berries, roots, and leaves. *Actaea* is from the Greek *aktea* ("elder"), referring to the similarity of baneberry leaves to those of elderberry. *Rubra* ("red") defines the primary color of the berries. Plants contain berberine and other toxic compounds; however, birds consume the berries and pass the seeds through their digestive tracts.

CANADIAN ANEMONE

Anemone canadensis
Buttercup family (Ranunculaceae)

Description: Perennial, with hairy stems up to 2' tall. Rounded leaves are deeply cut into 3–5 main lobes, sharply toothed along the margins, long stalked, and 3"–9" long; leaves may be sparsely hairy on either side. Flowering stalk bears pairs of opposite leafy bracts, which are stemless. Flowers have white sepals and lack petals, are about 2" wide, and bear numerous stamens (80–100) and a cluster of pistils. Fruit is a seed with sparse hairs.

Bloom season: Mid-spring to midsummer

Range/habitat: Across much of Canada and west-central and eastern United States in moist meadows and thickets, along river margins, and in swampy areas and lowland floodplains

Comments: *Anemone* may be from the Greek *anemos* ("wind"), referring to the seeds with long hairy plumes being scattered by the wind; another common name is windflower. *Canadensis* ("of Canada") refers to the type locality of the first specimen recorded for science. The roots contain astringent compounds that, along with the leaves, were used by North American Indigenous peoples as an astringent and treatment for wounds and cuts and as an eyewash. Also known as roundleaf anemone, for the shape of the leaves, and meadow anemone, for the plant's habitat preference.

JIM FOWLER

WESTERN PASQUEFLOWER

Anemone occidentalis
Buttercup family (Ranunculaceae)

Description: Perennial, with upright stout floral stems to 20" or more. Basal leaves are borne on long stalks; upper leaves are stemless. Both are divided 2–3 times into deep lobes. The single 1"- to 2"-long, white or purplish-tinged flowers have 5–8 oblong sepals and numerous stamens and pistils. Flowers lack petals. Fruit is a seed with long, silky hairs.

Bloom season: Late spring through summer

Range/habitat: Western North America on open, rocky slopes and in moist meadows

Comments: *Anemone* may be from the Greek *anemos* ("wind"), referring to the seeds with long hairy plumes being scattered by the wind; another common name is windflower. *Occidentalis* ("western") refers to the distribution of this plant. Another derivation of the name comes from Greek mythology. Flora, wife of Zephyr, god of the west wind, became jealous of Zephyr's passion for the nymph Anemone. Flora turned Anemone into a flower that would open when caressed by Boreas, god of the north wind. A colloquial name for the plant is hippie on a stick, for the long-haired appearance of the seed hairs.

WOOD ANEMONE

Anemone quinquefolia
Buttercup family (Ranunculaceae)

Description: Perennial, low growing, 3"–8" tall; from rhizomes that are white or black. A single rounded basal leaf arises from the rhizome on a long petiole, with lobed leaflets that may give the appearance of 5 leaflets, each about 1¾" long. Leaflets may be toothed or lobed. A separate stalk bears a whorl of 3 compound leaves, each 2–3 times divided, that subtend the single flower stalk; these leaflets are also toothed along the margins and may be lobed again. A single white to pink flower is borne on an elongated stalk; flowers are 1" wide, lack petals, and have 5 (4–9) petallike sepals. Fruit is a beaked seed, with or without hairs.

Bloom season: Early spring to early summer

Range/habitat: Eastern North America in moist thickets, along stream banks, and in open forests

Comments: *Anemone* may be from the Greek *anemos* ("wind"), referring to the seeds with long hairy plumes being scattered by the wind; another common name is windflower. *Quinquefolia* ("five-leaved") refers to the number of lobes on some of the leaves; generally there are 3 leaflets. The plants contain toxic and irritating glycosides.

STEVE R. TURNER

WHITE MARSH-MARIGOLD
Caltha leptosepala
Buttercup family (Ranunculaceae)

Description: Perennial; stems smooth and fleshy, growing 4"–16" tall. Basal leaves are oval to oblong, longer than broad, and somewhat arrowhead-shaped at the base. White or greenish flowers, 1"–2" wide, are borne 1–2 per stem. Outside of the flower is often tinged with blue. Fruit is a dry capsule that splits open.

Bloom season: Summer

Range/habitat: Western North America in wet meadows, seeps, and along stream banks

Comments: *Caltha* ("cup" or "goblet") describes the shape of the flower. *Leptosepala* ("with slender sepals") describes the sepals. Native Alaskan peoples ate the leaves and flower buds, as well as the plant's slender white roots. Beetles are common pollinators of these flowers. Elk browse on the leaves, which contain toxic alkaloids, without harm; hence cowslip and elk's lip are two other common names.

WESTERN VIRGIN'S BOWER
Clematis ligusticifolia
Buttercup family (Ranunculaceae)

Description: Perennial vine; stems may reach 30' long. Compound leaves have 3–7 lance- or egg-shaped leaflets that may be toothed along the edges. Small, inconspicuous flowers have white sepals but lack petals; may be few to many per flat-topped cluster. Fruit is a seed with a long, hairy tail.

Bloom season: Summer

Range/habitat: Western North America in moist soils in creek bottoms, along streamside thickets, and in coniferous woodlands

Comments: *Clematis* is the Greek name of a climbing plant. *Ligusticifolia* ("with leaves like *Ligusticum*) refers to the leaves resembling those of another genus. Leaves have a very peppery flavor and were used as a pepper substitute by pioneers. Native Americans chewed the leaves to treat colds or sore throats. Folklore suggests that the crushed roots were used like a "snuff" and inserted into a horse's nostrils to revive the horse. Another common name is old man's beard, for the seed's long, silky plumes.

STEVE R. TURNER

SHARP-LOBED HEPATICA

Hepatica acutiloba
Buttercup family (Ranunculaceae)

Description: Perennial, 2"–6" tall. Basal leaves are borne on a 6"-long hairy stalk, are about 3" long and palmately divided into 3 egg-shaped lobes that end in a pointed tip. Leaves may be mottled-brown looking prior to the flower opening, then turning green with white mottling. Single white to lavender flower, ½"–1" wide, is borne at the end of a 3"- to 4"-long, hairless stalk and has 6 (5–12) petallike sepals with numerous stamens surrounding a cluster of greenish pistils. Below the flowers are 3, ⅓"-long, lance- to egg-shaped hairy bracts. Fruit is a seed.

Bloom season: Early to late spring

Range/habitat: Central and eastern North America in deciduous woods; often in calcareous soils

Comments: *Hepatica* is from the Greek *hepar* ("liver") because the three-lobed leaf resembles the shape of the human liver. *Acutiloba* ("sharp-lobed") refers to the leaf tips. Round-lobed hepatica (*H. americana*) is similar but has rounded leaf tips. Both *Hepatica* species bloom early in spring. Another common name is liverleaf, for the mottled coloration or general outline of the leaves, which will overwinter.

STEVE R. TURNER

GOLDENSEAL

Hydrastis canadensis
Buttercup family (Ranunculaceae)

Description: Perennial, 9"–15" tall, with purplish stems. One or 2 hairy, maple leaf–shaped basal leaves are up to 8" wide and cut into 5–7 lobes. Flowering stalk arises from the base of an upper stalkless leaf and bears a single greenish-white to greenish-yellow, ½"-wide flower. Flower lacks petals, and the 3 petallike sepals fall off when the plant flowers. The prominent part of the flower is the numerous, about 50, white stamens. Fruit is a cluster of scarlet berries, each bearing 1 or 2 seeds.

Bloom season: Spring

Range/habitat: Eastern North America in wooded slopes or moist woods

Comments: The derivation of *Hydrastis* is perhaps from the Greek *hydor* ("water") and *drao* ("to act"), which may be in reference to the plant's use as a diuretic. *Canadensis* ("of Canada") refers to the distribution. The plant may reproduce by rhizomes, as it takes 4–5 years for a plant to reach sexual maturity, the point at which it flowers. The common name is for the yellow root's golden-colored sap, which was used by Native Americans and early settlers as a dye. The plant contains a variety of alkaloids, such as hydrastine and berberine, that have been used medicinally to treat various ailments; however, the alkaloids are toxic if taken in large doses.

STEVE R. TURNER

GOAT'S BEARD

Aruncus dioicus

Rose family (Rosaceae)

Description: Perennial, with hairless stems, 3'–6' tall. Lower leaves are 3 times compound; segments are ovate, toothed along the margins, and the tips are pointed. Upper leaves are smaller and 1–2 times compound. Tiny, white flowers are borne in long plumes of male and female flowers borne on separate plants. Male flowers have numerous stamens; female flowers have 3 pistils. Fruit is a narrow pod.

Bloom season: Late spring to midsummer

Range/habitat: Western and eastern North America along streams, forest edges, and roadsides in moist soils

Comments: *Aruncus* is from the Greek *aryngos* ("a goat's beard") and refers to the plumes of flowers resembling a goat's beard. *Dioicus* ("two homes") refers to the male and female flowers being borne on separate plants. Northwest tribes used the plants to treat blood ailments, as a diuretic, or to treat smallpox or sore throats. Another common name is bride's feathers, for the lacy-looking appearance of the flowers.

VIRGINIA STRAWBERRY

Fragaria virginiana

Rose family (Rosaceae)

Description: Perennial, low growing, 2"–7" tall; spreads by runners. Compound leaves are borne on hairy stems and divided into 3 ovate leaflets that are 2½" long with toothed margins. Upper surface of the leaf is dark green; the lower surface is variously hairy. A hairy flower stalk bears a small cluster of 4–6 white, 5-petaled flowers that are ½"–¾" wide and bear numerous stamens. Fruit is a red berry, about ½" wide.

Bloom season: Spring to early summer; may bloom a second time in the fall

Range/habitat: Eastern North America in fields, prairies, meadows, and dry open woodlands

Comments: *Fragaria* ("fragrant") may be for the sweet aroma of the fruits. *Virginiana* ("of Virginia") refers to the distribution and type locality of the plant. The plants may root along the runners like cultivated strawberries and often cover large patches. The common name strawberry comes from cultivated plants being mulched with straw to protect the edible fruits. A variety of insects, including bees, flies, and butterflies, pollinate the flowers. Coastal strawberry (*F. chiloensis*) grows along coastal areas from the Pacific Northwest to South America.

STEVE R. TURNER

WHITE AVENS

Geum canadense

Rose family (Rosaceae)

Description: Perennial, 1½'–3' tall. Lower leaves are brown, coarsely hairy, and compound with 3 leaflets; upper leaves are green and lightly hairy. Leaves are up to 4" long; the leaflets are lance- to oval-shaped, toothed along the margins, and divided into 3 deep lobes. Basal leaves are long stalked, about 6" long, and pinnately compound, with more than 3 leaflets that are coarsely toothed and hairy. White flowers (1–3) are borne in small clusters on hairy stalks about 3" long. The ½"-wide flowers have 5 petals, with 5 triangular-shaped sepals interspersed between the petals, and numerous stamens. Fruit is a seed with a hooked tip.

Bloom season: Summer

Range/habitat: Widespread across much of North America in thickets, forest edges, and shady woods

Comments: *Geum* is the Latin name for the plant. *Canadense* ("of Canada") refers to the plant's distribution. A variety of insects pollinate the flowers; the hooked seeds snag the fur of passing animals to aid in dispersal. There is a wide degree of variation between the basal leaves, which appear in winter, and the stem leaves later in the growing season. Pollinators include bees, wasps, and flies.

JIM FOWLER

BOWMAN'S ROOT

Gillenia trifoliata

Rose family (Rosaceae)

Description: Perennial, somewhat bushy; 2'–3' tall, with reddish stems. Leaves are divided into 3 segments that are narrow, oblong to lance-shaped, 1½"–4" long, and toothed along the margins. Leaves turn red in the fall. White, starlike flowers are 1"–1½" wide, with 5 linear petals and 5 sepals, which form a reddish calyx at the base of the petals.

Bloom season: Late spring to early summer

Range/habitat: Eastern North America in open forests, moist forests, and along roadsides

Comments: *Gillenia* honors Arnold Gillenius, a seventeenth-century German botanist and physician. *Trifoliata* ("three-parted") refers to the leaves. Native Americans used the powdered root as a laxative. Also known as Indian physic.

PARTRIDGE-FOOT

Luetkea pectinata
Rose family (Rosaceae)

Description: Small, mat-forming perennial, 2"–7" tall. Densely clustered basal leaves are finely divided and fan-shaped. Stem leaves may be absent. Upright flowering stems bear dense clusters of tiny, white to cream flowers. Fruit is a small capsule that splits open at maturity.

Bloom season: Summer

Range/habitat: Western North America in moist or shady sites in meadows and on rocky slopes

Comments: *Luetkea* honors Count F. P. Lütke (1797–1882), a German-Russian explorer and sea captain who mapped the coastline of Alaska during an expedition in 1826–1829. *Pectinata* ("like the teeth of a comb") describes the finely divided leaf. The common name refers to the outline of the leaf, which resembles the track of a partridge or ptarmigan. Also called Alaska spirea, for its northern distribution and spirea-like flowers.

ROCKY MOUNTAIN ROCKMAT

Petrophytum caespitosum
Rose family (Rosaceae)

Description: Mat-forming perennial, up to 3' or more across. Spatula- or inversely lance-shaped leaves, ⅛"–¼" long, have long, straight hairs on one or both sides or are smooth. Flowers are arranged in dense clusters at the end of a short stem; tiny petals are white. Fruit is a small capsule.

Bloom season: Midsummer to fall

Range/habitat: Western United States in rocky outcrops from sagebrush to alpine communities

Comments: *Petrophytum* ("rock plant") refers to the plant's habit of growing on rock surfaces or clinging to rock walls. *Caespitosum* ("low growing" or "growing in clumps") refers to the plant's stature. At times, the rock mats become so thick that even if the rock substrate below the plant gives way, the mat stays intact.

DAVID LEGROS

PARTRIDGE-BERRY

Mitchella repens
Madder family (Rubiaceae)

Description: Perennial, with trailing stems, up to 2" tall. Stems are 6"–12" long, overlapping to form dense mats. Evergreen leaves are round, ¾" wide, and glossy green with a white or yellowish mid-vein. White funnel-shaped flowers arise in pairs and are 4-lobed, hairy, and ½" long; petals may have a slight pinkish cast. Fruit is a reddish berry.

Bloom season: Mid-spring to midsummer

Range/habitat: Eastern North America in sandy sites around bogs, stream banks, bluffs, and moist woods

Comments: *Mitchella* honors John Mitchell (1711–1768), a Virginia physician who corresponded with Linnaeus, who named the plant in his honor. *Repens* ("creeping") describes the stature of this plant. The tasteless scarlet berries attract birds, especially partridges, but other wildlife species also consume the fruits. Native women brewed a tea from the leaves to aid in childbirth. The flowers are in two forms to prevent self-fertilization, and both must be pollinated to form a single berry.

BASTARD TOADFLAX

Comandra umbellata
Toadflax family (Santalaceae)

Description: Semiparasitic perennial; stems erect and 3"–13" tall. Linear, lance-shaped or narrowly elliptical leaves are ⅜"–1½" long and smooth. Flowers lack petals, have 5 whitish-green sepals, and are arranged in flat-topped clusters. Fruit is 1-seeded with a purplish or brownish fleshy coating.

Bloom season: Early spring to midsummer

Range/habitat: Western North America in shrublands, sagebrush steppe, woodlands, and roadsides

Comments: *Comandra* is from the Greek *home* ("hair") and *andros* ("man"), referring to the stamens, which are hairy at the base. *Umbellata* ("umbel-like") refers to the flat-topped flower clusters. Known to parasitize more than 200 plant species, including some in the oak (*Fagus*), rose (*Rosa*), and Goldenrod (*Solidago*) genera. The plant can photosynthesize to produce some of its own energy. This plant is also an alternate host for hard pine rust disease. The flowers resemble those in the toadflax genus (*Thesium*). The small edible fruits have a sweet taste.

CHUCK TAGUE

LIZARD'S TAIL

Saururus cernuus

Lizard's-tail family (Saururaceae)

Description: Perennial, up to 4' tall. Young stems may be hairy, becoming smooth with age. Heart- to lance-shaped leaves with a heart-shaped base are 3"–6" long. Long, slender spikes, 6"–12" long, of tiny, ¼"-long "flowers" that taper and droop at the tip arise from a leaf axil. The "flowers" lack sepals and petals but have 6–8 stamens and several carpels. Fruits are green and warty.

Bloom season: Summer to early fall

Range/habitat: Eastern North America and south to Texas in marshes, stream banks, swampy woods, and sloughs

Comments: *Saururus* is from the Greek *sauros* ("lizard") and *oura* ("tail"), referring to the white flowering spike that resembles a lizard's tail. *Cernuus* ("drooping") refers to the flowering spike. The plant has been used medicinally to treat swelling, inflammation, and fevers; however, the leaves are toxic. The roots, leaves, and flowers have a citrus aroma. Also known as breastweed or water dragon.

MATT BERGER

ELMERA

Elmera racemosa

Saxifrage family (Saxifragaceae)

Description: Perennial, 4"–10" tall. Basal leaves are round, 1"–1¾" wide, long stemmed, and palmately lobed. Margins are lobed or with rounded teeth. Flowers are borne along an elongated spike in loose clusters; sepals are yellowish white and very short, cup-shaped but hairy. The 5–10 white petals have 3–7 shallow lobes and are narrow at the bottom. Fruit is a capsule.

Bloom season: Late spring to late summer

Range/habitat: Pacific Northwest and British Columbia in rocky outcrops, crevices, or rocky slopes

Comments: *Elmera* honors Adolph D. A. Elmer (1870–1942), an American plant collector and botanist who collected widely in the western United States. *Racemosa* ("in a raceme") defines the arrangement of the flowers. Also known as yellow coralbells.

POKER ALUMROOT

Heuchera cylindrica
Saxifrage family (Saxifragaceae)

Description: Perennial. Mat-forming basal leaves are rounded to kidney-shaped, fleshy, and have 5–7 lobes that are toothed. Leaf blades are ½"–2½" long and borne on smooth or hairy stems, 1"–4" tall. Leafless flowering stems rise above the basal leaves, 6"–24" or more in height, depending on the growing site. Small white or pinkish cup-shaped flowers are borne in tight clusters; each flower is made up of 5 sepals and short stamens. Fruit is a seed.

Bloom season: Mid-spring through summer

Range/habitat: Western North America on rocky ledges or cliff faces and in rocky woods

Comments: *Heuchera* is for Johann Heinrich von Heucher (1677–1747), a German physician, botanist, and herbal expert. *Cylindrica* ("roll" or "cylinder") is in reference to the flower clusters, which are cylindrical or "poker"-shaped in outline. Also called lava alumroot, because this plant often grows in pockets in lava flows.

HENRIK KIBAK

BULBOUS WOODLAND STAR

Lithophragma glabrum
Saxifrage family (Saxifragaceae)

Description: Perennial. Deeply divided basal leaves may have 5 linear lobes or be divided into 5 toothed leaflets. Reddish bulblets form above bracts that are next to the flowers or along the stem. Flowers (1–7) are borne along a leafless, reddish stalk that has glandular hairs along the upper portion. White or light pink starlike flowers have 4 or 5 deeply divided petals, ¼" long. Fruit is a seed with small spines.

Bloom season: Early spring to late summer

Range/habitat: Western North America in grasslands, rocky meadows, coastal bluffs, and open woodlands, from sea level to 12,000'

Comments: *Lithophragma* is from the Greek *lithos* ("stone") and *phragma* ("fence"). *Glabrum* ("smooth") may refer to the leaves, although the leaves may have glandular hairs. The plants often reproduce asexually from the bulblets, which drop off and root. May grow alongside two other western species: slender woodland star (*L. tenellum*) and smallflower woodland star (*L. parviflorum*).

STEVE R. TURNER

EARLY SAXIFRAGE

Micranthes virginiensis
Saxifrage family (Saxifragaceae)

Description: Perennial, plants 2"–15¾" tall. Basal rosette of elliptical or egg-shaped leaves; leaves are 1⅛"–2¾" long, bluish green, and toothed along the margins. Hairy floral stalk rises up from the basal leaves to about 12"–18" tall and bears a cluster of 30 or more white, 5-petaled (sometimes 6-petaled) flowers; petals are separate, not fused, and about ¼" long. Fruit is a capsule.

Bloom season: Mid-spring to early summer

Range/habitat: Eastern and central North America in rocky fields, cliffs, rocky outcrops, along stream banks, and in dry woods

Comments: *Micranthes* is from the Greek *micro* ("small") and *anthos* ("flower"), referring to the tiny flowers in this genus. *Virginiensis* ("of Virginia") refers to the plant's distribution. The hairy stems may deter ground pollinators from accessing the flowers. Also known as Virginia saxifrage, lungwort, or sweet Wilson. The name saxifrage means "rock-breaker" and refers to the plants growing on rocky or stony surfaces.

NORTHWESTERN SAXIFRAGE

Saxifraga integrifolia
Saxifrage family (Saxifragaceae)

Description: Perennial, often 4"–8" long, but may reach 16" tall. Reddish stems may be hairy. Basal leaves are reddish and densely hairy below, variable in shape, and fringed with fine hairs on the edges. Flowers are borne in small clusters at the tip of a long stem; individual flowers have white or greenish petals and are ⅛"–¼" wide. Seeds have small ridges.

Bloom season: Early to mid-spring

Range/habitat: Western North America in grassy areas, along stream banks, and in subalpine meadows

Comments: *Saxifraga* is from the Latin *saxum* ("a rock") and *frango* ("to break"). *Integrifolia* ("entire leaves") refers to the smooth leaf edges. More than 20 species of saxifrage occur in the Pacific Northwest. Blooms early in the spring and often in profusion.

WESTERN JIMSONWEED

Datura wrightii
Potato family (Solanaceae)

Description: Annual or perennial, often in rounded clumps, 1'–4½' tall and covered with dense, fine gray hairs. Large leaves, 2"–10" long, have egg-shaped blades, are toothed or wavy along the margins, with short hairs and tapering to a point. Green sepals have lance-shaped lobes. Whitish to violet trumpet-shaped corollas are 5"–9" long and about as wide. Golf ball–shaped fruit is covered with prickles.

Bloom season: Late spring to late summer

Range/habitat: Southwestern United States and northern Mexico in sandy soils in desert canyons, shrublands, and woodlands

Comments: *Datura* is from the Sanskrit *dhatūra* ("thorn-apple"), referring to the thorny fruits. *Wrightii* honors Charles Wright (1811–1885), an American botanist who first collected this plant for science in Texas in 1850. The large flowers open in the evening and are pollinated by night-flying creatures such as hawkmoths, moths, and bats. The plants are toxic; the seeds contain psychoactive alkaloids and were consumed for their hallucinogenic properties. Also known as sacred thorn-apple or sacred datura, the plant has been used to induce visions but has also led to deaths.

COYOTE TOBACCO

Nicotiana attenuata
Potato family (Solanaceae)

Description: Annual, 1'–3' tall, with sticky hairs on stems. Leaves are elliptical to broadly lance-shaped and ½"–4" long; upper leaves are more lance-shaped than the lower, basal leaves. Trumpet- or funnel-shaped flowers are 1"–2" long, white, with lobes flaring open at the tips. Fruit is a small capsule containing tiny seeds.

Bloom season: Summer

Range/habitat: Western North America from British Columbia to Texas and northern Mexico in a variety of sites

Comments: *Nicotiana* honors Jean Nicot de Villemain (1530–1600), a French ambassador to Portugal who is credited with introducing tobacco plants into France in the sixteenth century. *Attenuata* ("pointed") refers to the pointed sepal tips. The plants contain nicotine as a defense mechanism; Native Americans harvested and smoked this plant as a tobacco alternative.

ELEANOR DIETRICH

CAROLINA HORSENETTLE

Solanum carolinense
Nightshade family (Solanaceae)

Description: Perennial, up to 3' tall. Undersides of the leaves and stems have white or yellowish spines. Leaves are oval to elliptical-oblong shaped, 2½"–6½" long and up to 3" wide, covered with fine hairs, and coarsely toothed or irregularly lobed. White, 5-petaled flowers are starlike and have stamens with stout yellow anthers; some flowers may have lavender or purple petals. Fruits are tomato-like and yellow at maturity.

Bloom season: Mid-spring to fall

Range/habitat: Southeastern United States in disturbed sites, prairies, abandoned fields, and forest edges

Comments: The derivation of *Solanum* is a bit uncertain but comes from the classic Latin name for a related plant and may be from *sol* ("sun"). *Carolinense* ("of Carolina") refers to where the plant was first collected for science. When crushed, the leaves smell like potato. The fruits contain toxic solanine glycoalkaloids; however, some birds and mammals consume the fruits and are unaffected by the toxins. Bumblebees are common pollinators of these flowers.

SITKA VALERIAN

Valeriana sitchensis
Valerian family (Valerianaceae)

Description: Perennial. Square stems range 1'–3' (up to 5') tall and are mostly smooth and somewhat succulent. Two or more pairs of leaves arise oppositely at points along the flowering stem. Compound leaves have 3–7 oval or lance-shaped leaflets, with the end leaflet being the largest. Leaf margins are coarsely toothed. Tiny, white to pinkish flowers are arranged in 1"- to 3"-wide, flat-topped or hemispherical clusters at the top of the plant. Stamens rise high above the mouth of 5-petaled, fragrant flowers. Fruit is a seed with feathery hairs on top.

Bloom season: Late spring through summer

Range/habitat: Western North America in moist meadows, along stream banks, and in moist forests and subalpine forests

Comments: *Valeriana* is probably derived from the Latin word *valere* ("to be healthy or strong") and refers to the plant's medicinal qualities. *Sitchensis* ("of Sitka") refers to the location of the first specimen collected for science. The roots have a strong aroma that is a sharp contrast to the sweet-smelling flowers. Northwest tribes used the pounded roots as a poultice for cuts and wounds. The seed's hairs act like parachutes, helping in dispersal.

TOM LEBSACK

BEAKED CORN SALAD

Valerianella radiata
Valerian family (Valerianaceae)

Description: Winter annual, 4"–24" tall, with square stems that may have hairs. Leaves form a basal rosette in the fall. Stem leaves are opposite, oblong to inversely lance-shaped, up to 3" long, and clasp the stem; leaf margins may have a few teeth along the lower portion and are variously hairy. Small, white, 5-petaled flowers are borne in flat-topped clusters; each cluster is about ½"–1" wide, and individual flowers are about ⅛" wide. Fruit is oval-shaped and 3-chambered, although only 1 chamber bears a single seed.

Bloom season: Mid-spring to early summer

Range/habitat: Southern and southeastern United States in moist meadows, plains, fens, pastures, disturbed areas, and open floodplains in woods

Comments: *Valerianella* is a derivation of *Valeriana*, which this plant resembles. *Radiata* ("radiated") may refer to the spreading leaves. The common name refers to the young leaves being collected and used as a potherb. The plants often grow in profusion, but they may be overlooked prior to flowering.

CLAIRE WEISER

CANADA VIOLET

Viola canadensis
Violet family (Violaceae)

Description: Perennial, 8"–16" tall. Both basal and stem leaves exist; blades of both leaves are heart-shaped, borne on long stems, 2"–4" long, and are finely toothed or scalloped along the margins. Stem leaves have shorter petioles than the basal leaves. Flowers arise from the axils of upper leaves on short stems. White flowers, ½"–1" wide, have 5 petals with yellow centers; backs of the petals may have a purplish tinge. There is a short spur, and the lower petal has purplish veins; the 2 lateral petals have small tufts of hairs near the throat. Fruit is an oval-shaped capsule covered with fine hairs.

Bloom season: Mid-spring to midsummer

Range/habitat: Across Canada and portions of the central and eastern United States in shady woods

Comments: *Viola* is the Latin name for various sweet-smelling flowers, including violets. *Canadensis* ("of Canada") refers to the plant's distribution and type locality. These violets may also bloom a second time in the fall. Bumblebees and other bees are common pollinators attracted to the pollen and nectar sources of these flowers.

YELLOW FLOWERS

This section includes yellow, golden, and yellowish-orange flowers. Some flowers have mixed colors, especially members of the Sunflower family, where ray and disk flowers are often in different colors; they are included in this section.

PATRICK ALEXANDER

PLAIN'S SPRING-PARSLEY

Cymopterus glomeratus
Carrot family (Apiaceae)

Description: Perennial, low growing, with flowering stems 2¾"–9" tall. Leaves spread laterally along the ground and are 2–3 times divided into lobes that are also deeply divided. Leaf petioles are 1"–4" long; the blades are about as long. Small, sticky hairs on the leaves and stems are often coated with sand grains. Clusters of small, yellowish flowers are arranged at the ends of a leafless stem. A tiny bract subtends the entire flower cluster; a secondary bract that is deeply cut into sharp-pointed segments subtends an individual flower head. Tiny petals are yellow, white, or purplish. Fruit is a ¼"-long seed with crinkled wings that are somewhat corky.

Bloom season: Early spring to early summer

Range/habitat: Portions of western and central North America into northern Mexico in desert shrub, plains, sagebrush, woodlands, and mountain meadows

Comments: *Cymopterus* is from the Greek *cym* ("waved") and *pteron* ("wing"), referring to the wavy margins along the winged fruits. *Glomeratus* ("ball") refers to the flower clusters. The parsnip-like taproot can be eaten raw or cooked. These plants lack basal stems; leaves arise from the root crown. Also known as Fendler's biscuitroot, honoring August Fendler (1813–1883), an American botanist who first collected this plant for science in 1849.

NINE-LEAF DESERT PARSLEY

Lomatium triternatum
Carrot family (Apiaceae)

Description: Perennial, 1'–3' tall. Compound leaves are dissected into 3 segments that are further divided into narrow, linear segments, ½"–5" long. Yellow flowers are borne in small clusters arranged to form a larger flat-topped cluster; the cluster stalk, or "ray," is about 4" long. Fruit is a flattened seed with papery wings and ribs.

Bloom season: Mid-spring to midsummer

Range/habitat: Western North America in moist to dry sites in meadows and open slopes

Comments: *Lomatium* is from the Greek *loma* ("a border or edge") and refers to the dorsal ribs, or "wings," that adorn the seeds. *Triternatum* ("triply ternate") describes the highly divided leaves. The flowers attract a variety of insect pollinators, which are also important food items for developing sage-grouse chicks. Also known as nine-leaf biscuitroot for the edible roots, which may be several feet long.

WESTERN SWEET-CICELY

Osmorhiza occidentalis
Carrot Family (Apiaceae)

Description: Perennial, 1'–4' tall, on smooth or hairy stems. Leaves are divided 1–3 times; the lance- to egg-shaped leaflets are lobed and toothed along the margins. Entire leaf blade is 4"–8" long. Flowering stalks lack leaves and terminate in umbrellalike clusters of tiny, greenish-yellow flowers. Fruit is a naillike seed.

Bloom season: Late spring to early summer

Range/habitat: Western North America in moist meadows, mountain slopes, and forests

Comments: *Osmorhiza* is from *osme* ("odor") and *rhiza* ("root"), for the licorice-like aroma of the crushed roots. *Occidentalis* ("western") describes this plant's distribution. The crushed leaves also have a sweet aroma. Aniseroot (*O. longistylis*), a similar-looking species that grows in the east, has white flowers and a root that smells like anise.

STEVE R. TURNER

PRAIRIE PARSLEY

Polytaenia nuttallii
Carrot family (Apiaceae)

Description: Biennial or short-lived perennial; may exist as a basal rosette of leaves for several years. Leaf blades are up to 6" long and 5" wide and highly divided into leaflets or sub-leaflets. Leaf stems are up to 4" long and may have a lightly hairy covering. Flowering stalk, when it emerges, is 1½'–3' tall. Tiny yellow flowers are arranged in dome-shaped or flat-topped clusters, 1½"–3" wide, and made up of 10–15 smaller clusters (umbellets); each smaller cluster has 10–15 flowers that are about ⅛" wide and have 5 petals and 5 sepals. Main cluster does not have persistent bracts, but the smaller umbellets have a few threadlike bracts. Fruit is an oval, flattened yellow seed with narrow, thickened margins.

Bloom season: Mid-spring to early summer

Range/habitat: Central United States in prairies, plains, savannas, glades, and thin woods; rarely in disturbed soils

Comments: *Polytaenia* is from the Greek *poly* ("many") and *tainia* ("fillet"), referring to the several oil tubes located between the seed's wings. *Nuttallii* honors Thomas Nuttall (1786–1859), an English botanist who collected plants widely in the western United States. The plant is a larval host to black swallowtail (*Papilio polyxenes*) larvae. Also known as wild dill.

SIMON TONGE

WILD ALLAMANDA

Pentalinon luteum
Dogbane family (Apocynaceae)

Description: Perennial vine, up to 12' long. Egg-shaped leaves are opposite, up to 6" long, and rounded at the tip. Leaves and stems have a milky latex. Yellowish flowers have 5 unfused sepals and 5 yellowish petals that are fused at the base to form a tube with reddish marks on the inside. At the top, the petals overlap to create a pinwheel-like appearance. Tip of the stamens has a curly appendage. Fruit is a narrow capsule, and the seeds have tufts of hairs.

Bloom season: Mid-spring to fall

Range/habitat: Florida south into South America and the Caribbean in dunes, woodlands, and rocky coastal areas

Comments: *Pentalinon* means "five cord," referring to the 5 flower petals and twining or cord-like nature of the plant. *Luteum* ("yellow") refers to the flower color. The plant is toxic, causing burning of the mouth, diarrhea, and convulsions if consumed. The milky sap has been used in parts of its range to treat dermatitis. Also known as hammock viper's-tail and yellow mandevilla.

YELLOW SKUNK CABBAGE

Lysichiton americanus
Arum family (Araceae)

Description: Perennial, 1'–2½' tall. Huge, broadly lance-shaped or elliptical leaves grow in a basal rosette around a stout stem. Leaves may be 1'–4½' long and half again as wide. A bright, yellow hood partially surrounds the greenish-yellow flowering stalk (called a spadix) that bears numerous tiny, greenish-yellow flowers. Berrylike fruit is embedded in the flowering stalk.

Bloom season: Spring to early summer

Range/habitat: Western North America in wet areas such as swamps, bogs, moist forest areas, fens, and along river bottoms

Comments: *Lysichiton* is from the Greek *lysis* ("releasing") and *chiton* ("a cloak") and refers to the yellow hood that falls apart with age. *Americanus* ("from America") refers to the distribution of this North American species. The plant contains oxalate crystals, which cause irritation and burning in the mouth and throat. Indigenous peoples of the Pacific Northwest ate the leaves roasted or steamed. Pollinators include beetles, bees, and flies; slugs often "pirate" the pollen and aren't considered pollinators.

KATJA SCHULZ

HEARTLEAF ARNICA

Arnica cordifolia
Aster Family (Asteraceae)

Description: Perennial; stems and leaves are covered with fine hairs. Basal leaves are opposite, toothed along the margins, and have a deep notch at the bottom to form a heart-shaped base. Upper leaves are broadly lance-shaped and toothed on the margins. Long flowering stalk, 6"–20", rises above the leaves and bears a single yellow flowering head. The 2"- to 3"-wide flower head has both ray and disk flowers; rays have shallowly notched tips. Fruit is a seed with white hairs.

Bloom season: Late spring to midsummer

Range/habitat: Western North America and most of Canada in moist open woods

Comments: *Arnica* is from the Greek word *arnakis* ("lambskin") and refers to either the woolly bracts that subtend the flower heads or the leaves' hairy undersides. *Cordifolia* ("heart-leaf") refers to the leaf bases, although there is some variation in the overall leaf shape. Arnicas are unique in this family due to the opposite leaves. Poultices and tinctures made from dried leaves are used as a disinfectant and to treat muscle strains and bruises. The seeds form apomictically (without fertilization).

DESERT MARIGOLD

Baileya multiradiata
Aster family (Asteraceae)

Description: Annual, biennial, or short-lived perennial, 12"–24" tall; often growing in clumps. Leaves are covered with silver-white hairs and are pinnately lobed; the lobes are mostly linear. Flowering stalks are nearly leafless, 4"–12" tall, and bear brilliantly yellow flower heads, about 1"–2" wide. Heads have both ray flowers (34–55), which are 3 lobed at the tip, and disk flowers (100+). Bracts that subtend the flower heads are linear-lanceolate. Fruit is a seed with hairs (cypsela).

Bloom season: Spring through fall

Range/habitat: Southwestern United States and northern Mexico in sandy or gravelly soils in open deserts, mesa tops, hillsides, and roadsides

Comments: *Baileya* honors Jacob Whitman Bailey (1811–1857), an American naturalist and professor of chemistry, mineralogy, and geology at West Point. Bailey made improvements to microscopes and used these instruments to conduct microscopic investigations on extensive collections of algae. *Multiradiata* ("many spreading out rays") refers to the many ray petals of these flowers. The leaves are toxic to sheep and goats but may be eaten by horses and cattle. Marigold is from "Mary's Gold," which refers to the Virgin Mary.

CRAIG MARTIN

ARROWLEAF BALSAMROOT

Balsamorhiza sagittata
Aster family (Asteraceae)

Description: Clump-forming perennial, ½'–2' tall and about as wide. Large, arrow-shaped leaves may reach 10" in length and 6" wide. Leaf margins are smooth; leaves are soft-hairy. Flowering stalks arise from the base of the leaves and bear yellow flower heads, 2"–4" wide. Both ray and disk flowers are present. Fruit is a seed.

Bloom season: Late spring to midsummer

Range/habitat: Western North America in dry meadows, grasslands, sagebrush flats, mountain brush, and pine forests

Comments: *Balsamorhiza* ("balsam root") refers to the aroma of the roots, which resembles balsawood. *Sagittata* ("arrow-shaped") refers to the outline of the leaves. Native Americans harvested the young roots and shoots in spring as a food source and for medicinal properties; mature roots may reach to 6' deep. Also known as Okanagan sunflower, this is the floral emblem for Kelowna, British Columbia. Carey's balsamroot (*B. careyana*) is similar, grows in the Pacific Northwest, and has rough-textured leaves and smaller flower heads, although the two may hybridize, creating confusion about their identity.

GREEN-EYES

Berlandiera lyrata
Aster family (Asteraceae)

Description: Perennial, 1'–2' tall, mound-forming. Leaves feel velvety to the touch, are pinnately lobed with odd length lobes, and have a chocolate-like aroma. Leafless flowering stalks bear 1"- to 2"-wide flower heads with yellow ray flowers with red veins on the undersides and maroon disk flowers. The flowers bloom in the morning and fade by mid-morning. Fruit is a cuplike seed head.

Bloom season: Late spring to mid-fall; in frost-free areas the flowers may bloom year-round.

Range/habitat: South-central United States into Mexico in dry, rocky soils in grasslands and disturbed sites

Comments: *Berlandiera* is for Jean-Louis Berlandier (1805–1851), a French physician, naturalist, and anthropologist who collected plants in Mexico before traveling to Texas as part of the Mexican Boundary Commission in 1827–1828. *Lyrata* ("lyre-shaped leaves") refers to the leaf's resemblance to the curves of a lyre, or ancient harp. Also known as chocolate daisy because the ray flowers and stamens also have a chocolatey aroma when crushed. The common name green-eyes is from the green disk, which resembles an eye after the ray flowers drop off.

STEVE R. TURNER

NODDING BUR-MARIGOLD

Bidens cernua
Aster family (Asteraceae)

Description: Annual, ½'–3' tall. Opposite leaves are linear to lance-egg shaped, up to 5" long, and toothed along the margins. Stems end in single flower heads that are ½"–1½" wide and somewhat nodding when developing. Two set of bracts subtend the flower heads—the inner ones are pale yellow and tapered to blunt tips; the 6–10 outer ones are green, linear, and leaflike. About 8 yellow, oval to elliptical ray flowers surround a cluster of yellow disk flowers; the petallike ray flowers are sometimes absent. Fruit is a seed with 4 barbed awns.

Bloom season: Midsummer to early fall

Range/habitat: Widely distributed across Canada and the United States in swamps, bogs, marshes, edges of streams and rivers, floodplains, and moist disturbed sites

Comments: *Bidens* ("two toothed") refers to the 2 teeth, or projections, on the seed tip. *Cernua* ("nodding") refers to the drooping flower heads. The noncompound leaves and 4 awns on the seeds distinguish this tickseed from others. The flowers attract a wide variety of insect pollinators.

STEVE R. TURNER

DEVIL'S BEGGARTICKS

Bidens frondosa
Aster family (Asteraceae)

Description: Annual, 8"–36" tall with square, reddish stems, generally branching only in the upper portions. Pinnately compound leaves are divided into triangular or lance-shaped leaflets (3–5) that are 2⅓"–3½" long, toothed along the margins, and with a pointed tip. Upper leaves are simple. Flower heads are borne singly or several, have 8 bracts that subtend the heads, are about ¾" wide and composed of yellow to orangish disk flowers; ray flowers may or may not be present. Fruit is a flattened black seed with hornlike projections on the tip.

Bloom season: Midsummer to mid-fall

Range/habitat: Across much of North America in prairies, meadows, disturbed areas, open woods, and along stream banks

Comments: *Bidens* ("two toothed") refers to the 2 teeth, or projections, on the seed tip. *Frondosa* ("leafy") refers to the leafy nature of the plant. Three-lobed beggarticks (*B. tripartita*) is similar and has 3 awns on the seed tip. *B. frondosa* is also known as leafy tickseed or devil's pitchfork for the awns on the seeds, which enable the seeds to snag on animal fur to help with dispersal.

CHUCK TAGUE

SEA OXEYE

Borrichia frutescens
Aster family (Asteraceae)

Description: Perennial, up to 3' tall. Oval to lance-shaped leaves are fleshy, hairy, toothed near the base or smooth along the margins, and up to 4⅓" long. Round flower heads have spine-tipped bracts below the flower heads, which have 15–30 ray flowers surrounding a center of yellow disk flowers with black stamens. Fruit is a flattened, triangular-shaped seed.

Bloom season: Generally late spring through summer but may bloom year-round, depending on conditions

Range/habitat: Southeastern United States and Mexico in coastal areas in beaches, dunes, and brackish wetlands

Comments: *Borrichia* honors Ole Borch (1628–1690), a Danish scientist and physician. *Frutescens* ("shrub-like") refers to the woody stems and overall growth. As the flowers mature and fall off, they are replaced with a spiny bur-like cluster of seeds. Butterflies are common pollinators of these flowers. Also known as sea daisy or beach carnation due to the plant's coastal distribution.

JIM FOWLER

GREEN-AND-GOLD

Chrysogonum virginianum
Aster family (Asteraceae)

Description: Perennial, 6"–12" tall, but may spread up to 18" wide. Stems have glandular hairs, and the plant sends out long runners. Leaves are egg-shaped, opposite, about 3" long, and toothed along the margins. Star-shaped flower heads are 1½" wide and arise from the leaf axils about 8"–10" high. The 5 petals are rounded, slightly notched at the tip, and surround a yellow center of disk flowers.

Bloom season: Mid-spring to midsummer, but may continue into mid-fall

Range/habitat: Eastern United States in wooded areas

Comments: *Chrysogonum* is from the Greek *chrysos* ("gold") and *gonu* ("joint"), in reference to the flowers arising from the leaf axils. *Virginianum* ("of Virginia") refers to the distribution of the plant. Also known as green knees or goldenstar for the shape and color of the flowers.

LAUREN MCLAURIN

MARYLAND GOLDEN ASTER

Chrysopsis mariana

Aster family (Asteraceae)

Description: Perennial, 1'–2' tall. Basal leaves form a rosette of long-petioled and somewhat inversely lance-shaped woolly leaves, 2"–6" long, and partially toothed along the margins; upper leaves are mostly elliptical, alternate, and less hairy than the lower ones. The ¾"- to 1"-wide flowering heads are composed of bright yellow ray flowers (12–21) surrounding a small center of yellow disk flowers. Two rings of bracts subtend the flower heads. Fruit is a reddish-brown seed.

Bloom season: Late summer to early fall

Range/habitat: Eastern and central United States in sandy soils in sandhills, pinelands, and roadsides

Comments: *Chrysopsis* is from the Greek *khrosus* ("gold") and *opsis* ("resembling in appearance"), referring to the golden flowers. *Mariana* ("of Maryland") is a reference to the plant's distribution. Also known as silkgrass in reference to the stems with fine hairs; the plants become less hairy with age.

STEVE R. TURNER

LANCELEAF COREOPSIS

Coreopsis lanceolata

Aster family (Asteraceae)

Description: Perennial, 12"–30" tall, often growing in small clumps. Lance-shaped hairy leaves, 3"–6" long, form a basal rosette and are lobed higher up the stems; leaves are also arranged oppositely but may be alternate higher on the stem. Flower heads, 1"–2" wide, are composed of 8–10 yellow ray flowers surrounding a large center of yellow disk flowers; the ray flowers have 4 lobes or notches at the tip. Fruit is a seed.

Bloom season: Mid-spring to midsummer

Range/habitat: Eastern and south-central North America in sandy soils in prairies, meadows, savannas, and open woods

Comments: *Coreopsis* is from the Greek *koris* ("bug") and *opsis* ("appearance"), which, along with the common name, refer to the shape of the seed resembling a tick. *Lanceolata* ("lance-shaped") refers to the shape of the leaves. Various butterflies and bees are attracted to these flowers as pollinators; the flowers also attract predatory insects that prey on garden pest insects. Also known as prairie coreopsis.

NAKED-STEM SUNRAY

Enceliopsis nudicaulis
Aster family (Asteraceae)

Description: Perennial, 4"–24" tall, with leafless flowering stalks arising from a woody base. Stout stem has dense, white hairs. Broadly rounded to egg-shaped leaves have long petioles and are covered with fine, silvery hairs around the stem base. Solitary, coarse flower heads are ¾" high and more than 2" wide. Each head has 13–21 yellow ray flowers that surround a dense cluster of yellow disk flowers. Fruit is a wedge-shaped seed with silky hairs.

Bloom season: Mid-spring to midsummer

Range/habitat: Southwestern United States in deep, well-drained soils in shrublands, rocky bluffs, and open woodlands

Comments: *Enceliopsis* ("similar to *Encelia*") refers to the plant's resemblance to another genus in the Asteraceae family. *Nudicaulis* ("naked stem") refers to the leafless flowering stalk. The closely related noddinghead (*E. nutans*) has flat, rounded leaves and a solitary flower head that lacks ray flowers. Some Native Americans of the Intermountain West prepared a tea from the roots to treat intestinal disorders and a tea from the leaves to treat coughs and colds. Bees, flies, and wasps are attracted to the plants as pollinators, as the flowers produce large amounts of pollen and nectar.

OREGON SUNSHINE

Eriophyllum lanatum
Aster family (Asteraceae)

Description: Perennial, 4"–24" tall, arising from multiple stems. Stems, leaves, and flower-head bracts are covered with dense, white hairs. Leaves are variable in shape, opposite or alternately arranged, and have 3–7 lobes at the tip. A single flower head, 1"–2" wide, is borne on a long stalk and bears 7–15 yellow ray flowers that are about ¾" long. Ray flowers surround a cluster of darker yellow disk flowers. Fruit is a smooth seed.

Bloom season: Late spring to fall

Range/habitat: Western North America into northern Baja California in dry sites, meadows, grasslands, sagebrush flats, rock slopes, and open woodlands

Comments: *Eriophyllum* ("woolly leaf") and *lanatum* ("woolly") define the hairy nature of the plant. These plants may bloom in profusion if conditions are right, as the seeds may lie dormant for years. A wide variety of pollinators, including beetles, flies, bees, moths, and butterflies, are attracted to the flowers; a rare Oregon butterfly, the Fender's blue (*Icaricia icarioides fenderi*), utilizes the flowers for nectar. Also known as woolly sunflower.

CHRISTIAN GRENIER

GRASS-LEAVED GOLDENROD

Euthamia graminifolia
Aster family (Asteraceae)

Description: Perennial, 2'–3½' tall. Plants arise on slender stems with fine white hairs and sometimes are arranged in dense clusters, making the plant appear bushy. Linear leaves are alternate, up to 5" long, have smooth margins, and may bear some white hairs. Larger leaves have 3 conspicuous veins; smaller leaves have a single conspicuous vein. Flower heads are arranged in flat-topped to rounded clusters of 20–35 flowers; each flower is about ⅛" wide and made up of 3–10 disk flowers and 7–35 ray flowers. Fruit is a seed with fine hairs.

Bloom season: Midsummer to mid-fall

Range/habitat: Across much of North America except parts of the southwestern and southeastern United States in wetlands, prairies, dunes, lake edges, disturbed sites, and ditches

Comments: *Euthamia* is from *eu* ("good" or "well") and *thama* ("crowded"), referring to the branching pattern. *Graminifolia* ("with grasslike leaves") refers to the shape of the leaves. Attracts a wide variety of pollinators, including butterflies, bees, moths, beetles, and wasps. Entire flowering clusters may be up to 1' across. Great plains goldenrod (*E. gymnospermoides*) is similar but grows in drier sites, has fewer flowers per cluster, and has 1 prominent vein on the larger leaves. Also known as flat-topped goldentop.

STEVE R. TURNER

BLANKETFLOWER

Gaillardia pulchella
Aster family (Asteraceae)

Description: Annual or short-lived perennial, 1'–3' tall, with hairy stems. Linear to lance-shaped leaves are mostly basal, 1½"–3" long, alternate, and may be lobed, toothed, or smooth along the margins. The 1"- to 2"-wide flowering heads bear reddish ray flowers with yellow, 3-cleft tips surrounding a center of reddish-brown disk flowers; sometimes the ray flowers are all yellow. Fruit is a pyramid-shaped seed.

Bloom season: Mid-spring to early fall

Range/habitat: Southern and central United States and northern Mexico in sandy plains, savannas and desert areas; naturalized across a much wider range

Comments: *Gaillardia* honors M. Chaillard de Charentoreau, an eighteenth-century French magistrate and patron of botany. *Pulchella* ("beautiful") refers to the color and shape of the flowers. This plant is a larval host for the border patch butterfly (*Chlosyne lacinia*). Also known as Indian blanket or firewheel due to the color and appearance of the flower heads; this is Oklahoma's state wildflower.

JOHN POLITES

CURLYCUP GUMWEED

Grindelia squarrosa

Aster family (Asteraceae)

Description: Perennial or biennial, up to 40" tall, with stout stems that are much branched. Elliptical or oblong leaves are ½"–2¾" long, with rounded teeth along the margins, and are glandular dotted. Flower heads bear 12–40 yellow ray flowers (sometimes ray flowers are absent) surrounding a center of yellow disk flowers and are about 1" wide. Below the flower heads are 5 or 6 rows of recurved or hooked bracts, also covered in resin. Fruit is a seed with stiff awns.

Bloom season: Summer to early fall

Range/habitat: Across much of western and central North America in prairies, savannas, grasslands, woodlands, disturbed areas, and along stream banks

Comments: *Grindelia* honors David Hieronymus Grindel (1776–1836), a professor of botany in Riga, Latvia. *Squarrosa* ("with upright scales") refers to the sticky, hooked or curved bracts. The dried leaves were added to cigarettes to treat asthma, which seems like a contradiction. Native Americans used the leaves and flowers to make teas to treat bronchial ailments. The bitter leaves are avoided by livestock, allowing the plant to expand its distribution into fallow fields and overgrazed rangelands.

PATRICK ALEXANDER

BROOM SNAKEWEED

Gutierrezia sarothrae

Aster family (Asteraceae)

Description: Perennial, rounded subshrub to 3' tall; branches resemble an inverted broom. Two types of leaves are present—linear stem leaves, ½"–2½" long, and smaller ones that grow in small clusters between the stem and the amin leaves. Flower heads, up to ⅛" wide, are arranged in flat-topped clusters. There are 3–7 ray flowers surrounded by 3–8 yellow disk flowers. Fruit is a seed with scales.

Bloom season: Midsummer to mid-fall

Range/habitat: Central Canada and much of the western United States in disturbed sites, grasslands, shrublands, and open sites

Comments: *Gutierrezia* is for Pedro Gutiérrez, a nineteenth-century botanist of Madrid. *Sarothrae* is from the Greek *sarotan* ("broom") and refers to the broomlike appearance of the stems. The stems were collected in bunches and tied off to create a crude broom. This species is often considered a weed because of its distribution on disturbed landscapes, but it is an indicator species of overgrazed lands and is toxic to livestock. One plant may produce up to 9,000 seeds. Also known as matchweed for the narrow stems and flammable nature of the plant.

STEVE R. TURNER

PURPLEHEAD SNEEZEWEED

Helenium flexuosum
Aster family (Asteraceae)

Description: Perennial, 1'–3' tall and clump forming; stem is winged below due to the clasping leaf bases. Basal leaves are egg- to lance-shaped, alternately arranged, and up to 3" long. Flower heads are about 1" across, with 8–14 yellow ray flowers that are 3-lobed at the tip and a rounded center of brownish-purple disk flowers. Ray flowers arch slightly downward, and a single row of lance-shaped bracts subtends the flower head. Fruit is a bullet-shaped seed, about 1" long, with a crown of scales at the top.

Bloom season: Midsummer to fall

Range/habitat: Central and eastern United States in moist soil along streams, ditches, ponds, swamps, prairies, woodlands, and pastures

Comments: *Helenium* is from the Greek *helenion,* a name for a different plant that honored Helen of Troy. According to Greek mythology, these plants sprang up wherever Helen of Troy's tears hit the ground. *Flexuosum* ("tortured" or "zigzag") is uncertain. The dried leaves and flowers were used as snuff; hence the common name. Numerous flower heads, which may lack ray flowers, form on each plant and attract a wide variety of insect pollinators. Common sneezeweed (*H. autumnale*) is similar but has a rounded center of yellow disk flowers.

STEVE R. TURNER

MAXIMILLIAN DAISY

Helianthus maximiliani
Aster family (Asteraceae)

Description: Perennial, upright stems reaching 3'–10' tall; stems feel coarse from short hairs. Linear leaves, up to 12" long, are alternate, folded lengthwise, coarse-hairy, which makes them gray greenish, and slightly toothed or smooth along the margins. Upper leaves may be 2" long. Numerous flower heads arise from leaf axils or the ends of the main stems, with 15–19 yellow ray flowers surrounding a center of green to brownish disk flowers; heads are 2"–3" wide but may reach 5" wide. Fruit is a seed with awns.

Bloom season: Late summer to mid-fall

Range/habitat: Central North America in plains, prairies, glades, rocky bluffs, and disturbed sites

Comments: *Helianthus* is from the Greek *helios* ("sun") and *anthos* ("flower") for the shape and color of the flower heads resembling the sun. *Maximiliani* honors Prince Maximilian of Wied-Neuwied, Germany (1782–1867), an ethnologist and naturalist who led an expedition to Brazil in 1815–1817 and one to the Great Plains of the western United States in 1832–1834 to record native cultures and collect plant and insect specimens. Maximilian collected this plant during his expedition to the Great Plains. Butterflies are common late-season pollinators of these flowers; birds eat the seeds later in the fall.

STEVE R. TURNER

JERUSALEM ARTICHOKE

Helianthus tuberosus
Aster family (Asteraceae)

Description: Perennial, 6'–10' tall, with stout hairy stems. Leaves are alternate below and opposite above and have a rough, hairy texture. Larger leaves are broadly egg-shaped with pointed tips and up to 12" long; upper leaves are narrower and smaller. Flower heads are 2"–4" wide, have 10–20 yellow ray flowers, and more than 60 disk flowers in the center. Fruit is a seed.

Bloom season: Late summer to early fall

Range/habitat: Eastern North America in moist thickets, fields, disturbed areas, and along stream banks and forest edges

Comments: *Helianthus* is from the Greek *helios* ("sun") and *anthos* ("flower") for the shape and color of the flower heads resembling the sun. *Tuberosus* refers to the tuberous root, which Native Americans harvested and cooked for food and is still cultivated today. The roots may be eaten raw or cooked like potatoes. The common name has nothing to do with Jerusalem but is believed to be a deviation of the Italian *girasole* ("turning toward the sun"), which early Italian immigrants called this plant. The taste of the roots resembles artichokes; hence the common name origin. Also known as sunroot or sunchoke.

CRAIG MARTIN

CAMPHORWEED

Heterotheca subaxillaris
Aster family (Asteraceae)

Description: Annual, biennial, or short-lived perennial. From a basal rosette of arrow-shaped leaves, the hairy flowering stems rise up ½'–5' tall. Alternate leaves along the stem are 1"–3¼" long, broadly oblong or lance- to egg-shaped, and with wavy margins that are slightly toothed or entire; the margins also have fine upturned hairs. Flowering heads are 1¼"–2¼" wide and composed of 15–45 yellow ray flowers and 25–60 yellow disk flowers. Below the flower heads are linear-lanceolate-shaped floral bracts in several rows. Fruit is a seed. Ray flowers produce a triangular-shaped seed without hairs; disk flowers produce a more oblong seed with fine hairs.

Bloom season: Midsummer to mid-autumn

Range/habitat: Southern half of the United States into the Northeast in prairies, fields, edges of agricultural areas, and disturbed sites

Comments: *Heterotheca* is from *heteros* ("different") and *theke* ("ovary"), in reference to the different types of seeds that form from the disk and ray flowers. If conditions are right, disk flower seeds may germinate immediately; ray flower seeds require a dormancy period. Various species of bees are common pollinators of these flowers. The crushed leaves have a camphor-like smell.

STEVE R. TURNER

DWARF ALPINE GOLD

Hulsea nana
Aster family (Asteraceae)

Description: Perennial, low growing; stems up to 6" tall. Leaves are mostly basal, hairy, ½"–2½" long, and inversely lance-shaped with shallow lobes. Flower stalks bear a single, bell-shaped flower head with numerous yellow ray and disk flowers. Bracts that subtend the flower heads are numerous and in 2 or 3 rows. Fruit is a seed with stiff hairs.

Bloom season: Mid- to late summer

Range/habitat: Western United States in subalpine or alpine environments in meadows or pumice soils

Comments: *Hulsea* honors Dr. Gilbert White Hulse (1807–1883), a US Army surgeon and botanist. *Nana* ("dwarf") refers to the plant's low stature. These fragrant flowers may bloom in profusion and attract late-season butterflies, bees, wasps, and flies as pollinators.

TWO-FLOWERED CYNTHIA

Krigia biflora
Aster family (Asteraceae)

Description: Perennial, 8"–30" tall; stems have milky latex. Inversely lance-shaped leaves, wider at the tip than the base, are mostly basal, up to 10" long and up to 2"–3" wide, rounded or pointed at the tip, and may or may not have small teeth or lobes along the margins. Upper leaves are lance-shaped, toothless, and clasp the stem at the base. Flowering stalks, which may be smooth or have glandular hairs, bear 1–6 yellow-orange or orange-tinted flowering heads that are 1"–1½" wide, with 25–60 ray flowers and no disk flowers. There are often 2 flowering heads per stalk. Ray flowers have 5 lobes or teeth at the tip. Fruit is a seed with 20–35 fine hairs.

Bloom season: Late spring to late summer

Range/habitat: Portions of eastern and central North America and parts of the southwestern United States in prairies, meadows, sandy fields, roadsides, and open forests

Comments: *Krigia* honors David Krieg (1667–1713), a German physician and botanist who first collected this plant for science in Maryland. *Biflora* ("two-flowered") refers to the 2 flowering heads per stalk. Also known as orange dwarf dandelion due to the flower's tint and dandelion-like appearance.

STEVE R. TURNER

GOLDEN RAGWORT

Packera aurea
Aster family (Asteraceae)

Description: Perennial, 6"–30" tall. Plants often spread by rhizomes. Basal leaves are dark glossy green, heart-shaped, toothed along the margins, about 2" wide and 2" long, and may have a purplish tinge on the underside. Leaf stems are about as long as the leaves. Stem leaves are pinnately lobed and smaller than the basal ones and lack hairs. Flowers are arranged in flat-topped clusters; flower heads are 1" wide and have both yellow ray and disk flowers. Bracts that subtend the flower heads are purple-tinged and pointed. Fruit is a seed with hairs.

Bloom season: Early spring to midsummer

Range/habitat: Eastern North America in moist shady areas in swamps, stream banks, springs, rocky slopes, and low-elevation woodlands

Comments: *Packera* honors John Packer (1929–2019), a Canadian botanist and professor of botany at the University of Alberta who studied the origin and evolution of Arctic flora. *Aurea* ("golden yellow") refers to the flower color.

ELEANOR DIETRICH

NARROWLEAF SILKGRASS

Pityopsis graminifolia
Aster family (Asteraceae)

Description: Perennial, with stems up to 2½' tall. Grasslike lower leaves are 4"–12" long; upper stem leaves are smaller. Leaves are covered with fine hairs, giving them a silvery appearance. Flower heads, ½"–¾" wide, have yellow ray flowers surrounding a center of yellow disk flowers. Fruit is a seed with fine hairs.

Bloom season: Throughout the year, but normally late summer to mid-fall

Range/habitat: Southeastern United States in dry prairies and pine woodlands

Comments: *Pityopsis* is from the Greek *pitys* ("pine") and refers to the Greek nymph Pitys, who was turned into a pine tree by Boreas, god of the north wind, after a quarrel. The plant's association with pine woodlands inspired the name. *Graminifolia* ("grasslike leaves") defines the leaves. Native Americans collected the leaves to treat colds, fevers, and achy joints. The plant attracts a wide diversity of bees and butterflies as pollinators. Also known as silver-leaved aster.

KATHY RIGALL

CAROLINA FALSE DANDELION

Pyrrhopappus carolinianus
Aster family (Asteraceae)

Description: Annual, 6"–20" tall, with erect stems with a milky sap. Leaves are 2"–6" long, deeply lobed along the margins, linear to inversely lance-shaped, and borne on stems that are one-half the length of the leaf. Upper leaves are smaller and clasp the stem so that the base of the leaf has "ears." Singular flower heads are 1"–1½" wide and have numerous ray flowers that are toothed at the tip and may be brownish or purplish tinged below. Disk flowers are absent. Heads may be arranged in loose clusters. Bracts that subtend the flower heads are in several rows and have minute hairs. Fruit is a seed with hairs.

Bloom season: Spring to fall

Range/habitat: South-central and eastern United States in pastures, roadsides, and disturbed sites

Comments: *Pyrrhopappus* is from *pyrrhos* ("yellowish-red") and *pappos* ("pappus"), in reference to the fiery color of the seed and hairs. *Carolinianus* ("of Carolina") refers to the plant's distribution. Like dandelions, the fine hairs on the seeds act like parachutes, enabling the seeds to disperse. Also known as Carolina desert-chicory and Texas dandelion for the plant's distribution and dandelion-like appearance.

STEVE R. TURNER

BLACK-EYED SUSAN

Rudbeckia hirta
Aster family (Asteraceae)

Description: Annual or short-lived perennial, 1'–3' tall. Stems and leaves are covered with rough hairs. Basal leaves are lance-shaped and 3"–7" long. Stem leaves are oval-shaped, up to 4" long, alternate, have shorter stems than the lower leaves, and taper to a tip. Flower heads are 2"–3" wide, borne singularly on a leafless stalk, and have 20–30 yellowish to yellow-orange ray flowers surrounding a dome of dark brown or purplish-brown disk flowers. Fruit is a seed without hairs.

Bloom season: Mid-spring to fall

Range/habitat: Central United States in dry prairies, fields, pastures, and roadsides

Comments: *Rudbeckia* is for Olof Rudbeck the Younger (1660–1740), who appointed Carl von Linnaeus as botany lecturer and head of botany demonstrations at Uppsala University in Sweden. Olof Rudbeck the Elder (1630–1702) was founder of the Uppsala Botanic Garden. *Hirta* ("hairy") refers to the bristlelike hairs on the stems and leaves. Rough coneflower (*R. grandiflora*) also grows in this region and is 3'–5' tall. Numerous types of pollinators are attracted to these flowers.

KATIE BYERLY

CUTLEAF CONEFLOWER

Rudbeckia laciniata
Aster family (Asteraceae)

Description: Perennial; flowering stalks grow 3'–12' tall. Lower leaves, up to 12" long and just as wide, are borne on drooping stems and have 3–7 coarsely toothed lobes, which may be divided again. Lobes are elliptical to egg-shaped. Upper leaves are entire and smaller. Flowering heads, 3"–4" wide, have 6–12 yellow ray flowers that surround a center of greenish disk flowers. A row of egg-shaped bracts, which are hairy or smooth, subtends the flowering head. This center elongates and forms a brownish cone shape as the seeds mature. Fruits are seeds.

Bloom season: Early summer to early fall

Range/habitat: Eastern North America in moist, open woodlands, thickets, stream banks, and sloughs

Comments: *Rudbeckia* is for Olof Rudbeck the Younger (1660–1740), who appointed Carl von Linnaeus as botany lecturer and head of botany demonstrations at Uppsala University in Sweden. *Laciniata* ("torn" or "deeply cut") is in reference to the leaf divisions. The flowers attract a variety of insect pollinators, especially butterflies. Also known as tall coneflower or greenhead coneflower for the color of the developing disk flowers.

ROUGH MULE'S EARS

Scabrethia scabra
Aster family (Asteraceae)

Description: Perennial; often in dense, sprawling clumps, 1'–3' tall and as wide or wider. Stems and leaves are covered with rough, stiff hairs. Leaves are linear to elliptical, 1"–7" long. Flower heads, 1"–3" wide, are generally solitary at terminal ends of branches. Yellow elliptical rays (10–23) surround a dense cluster of yellow disk flowers. Fruit is a seed.

Bloom season: Mid-spring to midsummer

Range/habitat: Southwestern United States in shrublands, low woodlands, and ponderosa pine forests

Comments: *Scabrethia* ("rough") refers to the texture of the leaves. *Scabra* ("rough") also refers to the texture of the leaves. Formerly known as *Wyethia scabra*, *Wyethia* honored Nathaniel Wyeth (1802–1856), a Massachusetts businessman who led two overland expeditions to the Oregon Territory in 1832 and 1834. Botanist Thomas Nuttall and ornithologist John Kirk Townsend accompanied the second expedition, both collecting plants and birds new to science.

STEVE R. TURNER

ARROWLEAF GROUNDSEL

Senecio triangularis
Aster family (Asteraceae)

Description: Perennial; cluster of tall stems rising to 5' tall. Arrow-shaped leaves are 2"–8" long, alternately arranged along the stem, strongly toothed, and more or less hairless. Flower heads bear few to 8 ray flowers that are toothed at the tip; the 5–8 ray flowers surround a small but dense cluster of disk flowers. Fruit is a seed with fine hairs.

Bloom season: Late spring to midsummer

Range/habitat: Western North America in moist meadows, thickets, open forests, and along stream banks

Comments: *Senecio* is from the Latin *senex* ("old man") and refers to the white hairs atop the seeds. *Triangularis* ("triangle-shaped") refers to the shape of the leaves. The plant's flowers and seeds contain pyrrolizidine alkaloids, which are toxic to wildlife and livestock in large quantities; most animals graze on these plants prior to flowering.

PAINTED DESERT BAHIA

Silphium integrifolium
Aster family (Asteraceae)

Description: Perennial, generally 2'–3', but up to 6' tall. Opposite leaves (may be alternate or whorled) are variable, lance- to egg-shaped, up to 6" long and 2½" wide, with a rough texture; margins may or may not be hairy or toothed. Flower heads bear 15–35 yellow ray flowers surrounding a center of yellowish disk flowers and are 2"–3" wide. Bracts that subtend the flower heads are recurved. Fruit is a seed produced by the ray flowers only.

Bloom season: Summer to fall

Range/habitat: Central United States in prairies, savannas, rocky woods, and disturbed areas

Comments: *Silphium* is from the Greek *silphon*, referring to a North African resin-bearing plant. *Integrifolium* ("entire or uncut leaves") refers to the shape of the leaf. Native Americans harvested the root for use as a pain reliever. Also known as rosinweed for the sticky resinous sap that exudes from the crushed or broken stems.

STEVE R. TURNER

COMPASS PLANT

Silphium laciniatum
Aster family (Asteraceae)

Description: Perennial, 3'–12' tall, with a stout stem that exudes a sticky sap when broken. Basal leaves are deeply lobed, up to 2' long, and arranged in a north–south orientation. Stem leaves are smaller, clasp the stem, and are alternately arranged. Flower heads arise on long, hairy stalks, are 2"-5" wide, and have 20–30 yellow ray flowers that surround a center of yellow disk flowers. Fruit is a seed; disk flowers are sterile.

Bloom season: Summer to fall

Range/habitat: Central and eastern North America in dry prairies, savannas, roadsides, and disturbed sites

Comments: *Silphium* is from the Greek *silphon,* which refers to a North African resin-bearing plant. *Laciniatum* ("shredded" or "torn to pieces") refers to the divided leaves. The cut stem exudes a resin that was chewed like gum. Cup plant (*S. perfoliatum*) has square stems and opposite leaves that form a cup where they join the stem.

STEVE R. TURNER

TALL GOLDENROD

Solidago altissima
Aster family (Asteraceae)

Description: Perennial, 2'–4' tall, with rough hairy stems. Leaves are somewhat lance-shaped, 2"–6" long, broadest in the middle, and tapering to a tip and stalkless base; leaf edges are variable. Leaves are variously hairy—the upper surface is rough; the lower surface is hairier along the veins. Flowering heads are densely arranged along one side of the stalk, sometimes resembling a pyramid with more than 1,200 flowers per head. Three to four rows of lance-shaped bracts subtend the flower heads. Flowering heads are about ¼" wide, with 8–15 yellow ray flowers surrounding a small center of 3–6 yellow disk flowers. Fruit is a seed with fine bristles.

Bloom season: Late summer to mid-fall

Range/habitat: Central and eastern North America and northern Mexico in open fields, meadows, prairies, roadsides, and woodland edges

Comments: *Solidago* is from the Latin *solido* ("to make whole or heal") for the medicinal properties of the plants. *Altissima* ("tall") refers to the stature of this plant, which is also known as late goldenrod due to its flowering time. The stems often bear numerous insect galls. Canadian goldenrod (*S. canadensis*) is a similar-looking species, with smaller flower-head bracts and hairs only along the leaf veins on the undersides. Goldenrods also spread by rhizomes and may cover large patches of ground.

CHUCK TAGUE

SEASIDE GOLDENROD

Solidago sempervirens
Aster family (Asteraceae)

Description: Perennial, 2'–8' tall. Basal rosette of egg- to inversely lance-shaped leaves that are evergreen and somewhat fleshy, hairless and toothless, and up to 16" long. Upper leaves are smaller and alternate. Stems bear terminal clusters and numerous flowering side branches; most of the flowering stalks are somewhat curved or reflexed. Dense clusters of ¼"-wide dark yellow flowers are arranged along one side of the stalks and include both ray and disk flowers. Fruit is a seed with fine hairs.

Bloom season: Midsummer to late fall, but may bloom throughout the year

Range/habitat: Eastern North America, the Caribbean, and the southern United States from Florida to Texas in coastal dunes, beaches, marshes, and pine woodlands

Comments: *Solidago* ("becoming whole or strengthen") refers to the medicinal properties of the plant. Many species have been used by Native Americans, the ancient Greeks, and modern-day herbalists for treating various ailments. *Sempervirens* ("evergreen") refers to the persistent leaves. The salt tolerance of this plant enables it to inhabit very brackish habitats. Flowers attract a wide variety of pollinators.

SCAPEPOSE GREENTHREAD

Thelesperma subnudum
Aster family (Asteraceae)

Description: Perennial, stems 1"–20" tall. Leaves are mainly basal, opposite, ¾"–3" long, divided to the midrib, with the lobes linear. Flower stalks are leafless; the heads are ¼"–¾" wide, with small outer bracts below the heads that are lance-shaped and bent outward. There are often 8 yellow ray flowers (sometimes none), ⅜"–1⅛" long and lobed at the tip, surrounding a center of yellow disk flowers. Fruit is a seed with a pointed tip.

Bloom season: Mid-spring to midsummer

Range/habitat: Central North America in open areas, rocky slopes, and woodlands

Comments: *Thelesperma* is from the Greek *thele* ("nipple") and *sperma* ("seed"), referring to the small, pointed projection at the tip of the seeds. *Subnudum* ("partially naked") refers to the flowering stalks, which bear leaves only along the lower portion. Also known as Navajo tea, for a tea brewed from the dried flowers and young leaves, or border goldthread.

TOM LEBSACK

GOLDEN CROWSBEARD

Verbesina encelioides
Aster family (Asteraceae)

Description: Annual, 2'–5' tall. Triangular gray-green leaves are alternate or opposite, have toothed margins, and are up to 4" long. Flower heads are up to 2" wide, have 8–15 ray flowers that have 2 longitudinal grooves and are 3-notched at the tip. Ray flowers surround a center of 80–150 yellow disk flowers. Bracts that subtend the flower heads are in 1–2 series and are lance- to oval-lance-shaped and recurved. Fruits are seeds.

Bloom season: Midsummer to early fall

Range/habitat: Widespread across parts of the United States and Mexico in fields, washes, and disturbed sites

Comments: *Verbesina* ("*Verbena*-like") refers to the leaves resembling those of verbenas. *Encelioides* ("like *Encelia*") refers to the plants being similar to those in the *Encelia* genus. The basal leaves are generally alternate; the upper leaves are opposite. This plant is a larval host for the sunflower patch butterfly (*Chlosyne lacinia*). The Navajo in the Southwest placed sprigs of flowers in their hatbands to be protected from lightning. The flowers attract a variety of moths and butterflies as pollinators. Also known as cowpen daisy or butter daisy.

STEVE R. TURNER

NARROWLEAF PUCCOON

Lithospermum incisum
Borage family (Boraginaceae)

Description: Perennial; low growing, but may reach 20" tall. Dark green linear leaves, up to 3" long, have small hairs pressed close to the surface, are alternate, and have margins that roll inward. Each trumpet-shaped yellow flower has a long, thin corolla tube that flares into 5 ruffled lobes. Flowers are about 1" wide, and the toothed lobes have fine hairs. Fruit is a hard white nutlet.

Bloom season: Early spring to early summer

Range/habitat: Western and central North America in rocky soils, prairies, savannas, shrublands, and woodlands

Comments: *Lithospermum* ("stone seed") refers to the hard seed. *Incisum* ("toothed") refers to the lobes of the flowers. Blood-stimulating teas were made from the stems, leaves, and roots. Though the flowers produce some seeds, later in the season the plant produces closed, self-fertilizing flowers in the lower axils that produce nutlets.

WESTERN PUCCOON

Lithospermum ruderale

Borage family (Boraginaceae)

Description: Perennial, 1'–2' tall; numerous stems arise from a woody root base. Upper leaves are linear and 1"–4" long; lower leaves are smaller. Small trumpet-shaped yellow flowers, ½" wide, are borne at the stem tips and nearly obscured by the upper leaves. Fruit is a hard nutlet.

Bloom season: Mid-spring to late summer

Range/habitat: Western North America in dry foothills, plains, sagebrush shrublands, woodlands, and disturbed sites

Comments: *Lithospermum* ("stone seed") refers to the hard nutlets of this genus. *Ruderale* ("rubbish") refers to the plant's growth habit in disturbed areas. Plains Indians harvested the roots as a food source and to make a remedy for treating respiratory illnesses. Puccoon is the Native American word for a dye plant; several species of *Lithospermum* yield a purplish dye. Also known as western stoneseed.

YELLOW CRYPTANTHA

Oreocarya flava

Borage family (Boraginaceae)

Description: Perennial, 4"–16" tall, often with clustered stems. Leaves, flowering stems, and sepals are covered with dense stiff hairs. Linear or inversely lance-shaped leaves are ¾"–3" long and mainly basal; leaves are opposite at the base and alternate along the stem. Dense clusters of tubular yellow flowers, ¼" wide, have 5 petals that flare open at the top of a short tube. Small, arching crests encircle the open mouth of the tube. Fruit is a nutlet.

Bloom season: Early spring to midsummer

Range/habitat: Southwestern United States in desert shrub, sagebrush, and pinyon-juniper woodlands

Comments: *Oreocarya* is from the Greek *oreos* ("mountain") and *karyon* ("a nut"), referring to the montane growing habit of many members of this genus. *Flava* ("yellow") describes the flower color. Seeds may stay dormant in the ground until conditions are favorable to germinate, and some seasons these plants may carpet large desert areas. Some botanists place this plant in the *Cryptantha* genus. Long-tongued bees are common pollinators of the flowers.

WESTERN WALLFLOWER

Erysimum asperum
Mustard family (Brassicaceae)

Description: Biennial or perennial, stems 1'–3' tall. Basal leaves are linear- or spatula-shaped, up to 4" long, widest in the middle, with several widely spaced teeth along the margins; leaves have Y-shaped hairs. Upper leaves are alternate, may have smooth or slightly toothed margins, and are covered with star-shaped hairs. Yellow to yellow-orange flowers have 4 petals and 6 stamens, are about ¾" wide, and arise in clusters from a leaf axil. Fruits are slender pods, up to 3" long.

Bloom season: Mid-spring to early summer

Range/habitat: Across Canada and the central United States in sandy or rocky soils in plains, prairies, and along roadsides and stream banks

Comments: *Erysimum* ("help or save") refers to the plant's reported beneficial uses as a poultice. *Asperum* ("rough") refers to the texture of the leaves. Northern wallflower (*E. capitatum*) grows in western North America and has dense clusters of 4-petaled yellow flowers.

TWINPOD

Physaria newberryi
Mustard family (Brassicaceae)

Description: Perennial, 1"–9" tall; low growing. Basal leaves, up to 3" long, have rounded, egg- or spatula-shaped blades with a small pointed tip. Leaves along the flowering stalk are smaller. Yellow flowers, each arising on a short stalk, are borne in tight clusters. Each flower has 4 spatula-shaped petals and is ¼"–½" wide. Inflated seedpod has a deep indentation between the halves.

Bloom season: Late winter to midsummer

Range/habitat: Southwestern United States in sandy or rocky soils in desert shrub, grasslands, or woodlands

Comments: *Physaria* is from the Greek *physa* ("bladder") and refers to the inflated seedpod. *Newberryi* honors John Strong Newberry (1822–1892), a professor of geology at Columbia University; he served on the Ives Expedition (1857–1858) to the southwestern United States and is credited as being the first geologist to visit the Grand Canyon. The common name refers to the 2 halves of the inflated seedpod.

PRINCE'S PLUME

Stanleya pinnata
Mustard family (Brassicaceae)

Description: Perennial, up to 6' tall, with long flowering stalks. Lower leaves are deeply dissected, ½"–6" long and ¾"–1½" wide. Upper leaves are narrowly lance-shaped or elliptical. Elongated flowering stalk bears clusters of lacy yellow flowers, ¾"–1½" long, with stamens that protrude above the 4 petals. Seedpods are long stalked, narrow, and up to 3" long.

Bloom season: Late spring to late summer

Range/habitat: Western United States in selenium-bearing soils in desert shrub or mountain brush communities

Comments: *Stanleya* is for Lord Edward Stanley (1755–1851), a president of the Linnaean and Zoological Societies in London. *Pinnata* ("pinnate") refers to the deeply dissected leaves. Presence of this species indicates selenium-bearing soil. It is toxic to livestock; however, native bighorn sheep forage on this plant with no ill effects. This plant is a larval host to the cabbage white butterfly (*Pieris rapae*). Becker's plume (*S. albescens*) grows in the southwestern United States but has white flowers.

BILLY AGUIRRE

BUCKHORN CHOLLA

Cylindropuntia acanthocarpa
Cactus family (Cactaceae)

Description: Perennial, shrub-like cactus, 3'–9' tall. Jointed stems, which are greater than 6" long, are green, cylindrical, and covered with straw-colored spines, 6–20 per cluster. Flowers are yellow-orange, pink, orange, or bronze, 2"–3" wide, with numerous stamens. Fruit is a spiny, fleshy fruit.

Bloom season: Late spring to early summer

Range/habitat: Southwestern United States in grasslands and shrublands

Comments: *Cylindropuntia* is from the Greek *cylindrus* or Latin *kylindros* ("cylinder") and *opuntia*, which refers to the ancient Greek city of Opus, where a spiny plant could sprout from the stems being planted. *Acanthocarpa* is from *akanthos* ("spines" or "thorns") and *carpos* ("fruit"), in reference to the spiny fruits. Teddy bear cholla (*C. bigelovii*) is another common Sonoran Desert cactus. Native tribes still harvest the cholla buds and roast them in pit-ovens; the flavor resembles cooked asparagus. Jointed stems that break off from the main plant may root. The branching stems resemble a buck's antlers; hence the common name.

STEVE R. TURNER

PRICKLY PEAR CACTUS

Opuntia polyacantha

Cactus family (Cactaceae)

Description: Perennial cactus, 4"–12" tall; clump forming up to 4' wide. Large, flattened pads are variable in shape and size, from egg-shaped to oval, and are 2"–8" long. Spines, ¼"–3" long and 5–11 per cluster, are white, flattened at the base, and straight or curved slightly downward. Flower color varies from yellow to pink or reddish. Bowl-shaped flowers are 2"–3½" wide, with numerous stamens and smooth petals. Pulpy fruits become reddish and spiny.

Bloom season: Late spring to midsummer

Range/habitat: Western and central United States in grasslands, shrublands, and dry rocky sites

Comments: *Opuntia* is a Greek name for a spiny plant that grew near the city of Opus, Greece. *Polyacantha* ("many spined") describes the numerous spines. The brilliantly colored flowers attract bees and beetles as pollinators; it is not uncommon to see numerous insects wading through the stamens inside the flower. The sweet, edible fruits are harvested and the spines carefully removed through burning or cutting.

YELLOW BEEPLANT

Cleome lutea

Caper family (Capparaceae)

Description: Annual, 1'–3' tall, with stout stems. Alternate palmately compound leaves have 3–7 elliptical or lance-shaped leaflets, ⅜"–2" long. Flowers have short stalks, 4 petals, and 6 to many stamens that protrude out of the flower. Flowers are arranged along an elongated stalk. Fruit is a capsule, ⅜"–1½" long, that hangs downward on a short stalk and bears black seeds.

Bloom season: Mid-spring to late summer

Range/habitat: Western United States in sandy soils or disturbed sites in desert shrub and grassland areas

Comments: *Cleome* is an ancient Greek name for another plant in the Mustard family. *Lutea* ("yellow") refers to the flower color. During springs with abundant moisture, fields of this beeplant may carpet the ground. Native Americans in the southwestern United States boiled the flowers to make a black pigment used on pottery. Golden spider flower (*C. platycarpa*) grows in portions of the northwestern United States and has gland-tipped leaves.

JIM FOWLER

LIMBER HONEYSUCKLE

Lonicera dioica

Honeysuckle family (Caprifoliaceae)

Description: Perennial vine, 3'–10' long. Leaves are opposite, 1½"–4½" long, smooth, and oval to egg-shaped; uppermost leaf below the flowers is fused and cuplike. Red, yellow, or pale orange tubular flowers are ¾" long, 2-lipped, and with long exserted yellow stamens. Fruit is a red berry.

Bloom season: Late spring to midsummer

Range/habitat: Across Canada and central and eastern United States in boggy areas, clearings, and moist sites in thickets and woodlands

Comments: *Lonicera* honors Adam Lonitzer (1528–1586), a German botanist and physician. *Dioica* is from a Greek word meaning "two houses," in reference to having male and female flowers on different plants. Hummingbirds are common pollinators of the flowers. Also known as glaucous honeysuckle or red honeysuckle for the color of the flowers and berries.

ERIK ERBES

PINEBARREN FROSTWEED

Crocanthemum corymbosum

Rock-Rose family (Cistaceae)

Description: Perennial, with hairy stems up to 8" tall; mound-forming. Evergreen leaves are elliptic or linear, 1"–1¾" long, with edges that roll under. Yellow shallow bowl-shaped flowers have 5 petals with squared tips, are about ½" wide, and have stamens with orange anthers. Flowers are arranged in a flat-topped cluster, opening from the outside first, then toward the inside. Fruit is a capsule.

Bloom season: Early spring to early summer

Range/habitat: Southeastern United States in sandhills, dunes, and woodlands

Comments: *Crocanthemum* is from the Greek *crocus* ("yellow") and *anthem* ("a flower"), in reference to the color of the flowers. *Corymbosum* is from the Greek *corymb* ("the top or cluster of flowers"), which refers to the terminal flower clusters. For most of the year the plants are low growing, forming mounds of vegetation that may be 2'–3' wide.

MATT BERGER

CANYON LIVE FOREVER

Dudleya cymosa

Stonecrop family (Crassulaceae)

Description: Perennial succulent; reddish-orange stems. Basal rosette, 2" tall and up to 11" wide, of gray-green hairy leaves that are broadly lance- to spoon-shaped with a tip. From the rosette arises a reddish flowering stalk, up to 1' tall, that bears a flat-topped cluster of flowers. Small yellowish-red, thimble-shaped flowers are borne at the top of a stem; petals are lance-shaped. Fruit is a seed.

Bloom season: Late spring

Range/habitat: California and Oregon in rocky areas and cliff faces from near sea level to over 9,000' in elevation

Comments: *Dudleya* honors William Russel Dudley (1849–1911), the first chair of the Stanford University botany department, whose plant collection there is one of the finest representing California flora. *Cymosa* is from a Greek word meaning "flowers arranged in a cyme," which refers to the arrangement of the flowers. These plants attract hummingbirds and butterflies as pollinators. The family name is derived from a photosynthetic pathway known as Crassulacean acid metabolism (CAM) used by various succulent desert plants, including cacti and stonecrops. The plants may reproduce by seeds or clones; hence the plants seem to live forever.

LAURA HOLLOWAY

LANCELEAF STONECROP

Sedum lanceolatum

Stonecrop family (Crassulaceae)

Description: Perennial succulent, 4"–5" tall. Basal leaves are linear to oval, pointed or knobby, up to 1⅛" long, alternate, and rounded; upper leaves are smaller and tend to drop off before the flowers open. Starlike flowers are borne in dense clusters at the ends of the stems. Yellow petals are ⅓" long, may be tinged with red, and have stamens with yellow anthers. Fruit is a tiny seed.

Bloom season: Late spring to late summer

Range/habitat: Western North America in rocky outcrops, gravelly sites, or on ledges from sea level to alpine elevations

Comments: *Sedum* is from the Latin *sedo* ("to sit"), in reference to the plant's habit of growing on rocky ledges or stone walls. *Lanceolatum* ("lance-shaped") refers to the shape of the petals. Besides seeds, the plants may spread by sections of the stem breaking off and rooting. This sedum is a host plant for the Rocky Mountain parnassian butterfly (*Parnassius smintheus*); the larvae ingest the plant's cyanogenic glycosides, which make the caterpillars unpalatable. However, if the larvae feed on the leaves in winter, the toxicity is fatal to them. Also known as spearleaf stonecrop for the shape of the leaves.

TOM LEBSACK

MELÓN LOCO

Apodanthera undulata

Gourd family (Cucurbitaceae)

Description: Perennial vine, with a massive taproot up to 8" wide. Stems are prostrate, up to 10' long, hairy, and may have tendrils for climbing. The round to kidney-shaped leaves are lobed, up to 6" wide, gray-green due to hairs, and have wavy or toothed margins. Trumpet- to funnel-shaped yellow flowers are either male or female (on the same plant), 5-lobed, average about 1½" long, and foul smelling. Male flowers (1–5) are in flat-topped clusters; female flowers (5–12) are arranged in bundles. Fruit is an egg-shaped gourd, about 4" long, with longitudinal ridges.

Bloom season: Early summer to early fall

Range/habitat: Southwestern United States, Arizona to Texas, and into Mexico and South America in dry river washes, plains, and mesas

Comments: *Apodanthera* is from the Greek *a* ("without"), *podos* ("foot"), and *anthera* ("anther"), referring to the sessile anthers. *Undulata* ("wavy") refers to the leaf margins. The stiff hairs on the leaves feel like sandpaper. The saying is "These taste so bad you'd have to be *loco* to eat one."

STEVE R. TURNER

BUFFALO GOURD

Cucurbita foetidissima

Gourd family (Cucurbitaceae)

Description: Perennial vine with several stems that may reach up to 20' long. Large triangular leaves are shallowly lobed or angled along the margins. Thick leaves are heart-shaped at the base and pointed at the tip. Funnel-shaped yellow flower, 3"–5" long, has an unpleasant scent. Fruit is a rounded gourd with stripes.

Bloom season: Early to late summer

Range/habitat: Central and southwestern United States and into northern Mexico in disturbed sites and wash bottoms

Comments: *Cucurbita* is from a Latin word for a type of gourd. *Foetidissima* ("fetid") refers to the strong odor of the crushed leaves, stems, or flowers. Native Americans used the gourds for water containers and as food; seeds have been found in archaeological digs in the Southwest. A biodiesel fuel has been made from the seeds.

PINESAP

Monotropa hypopitys
Heather family (Ericaceae)

Description: Myco-heterotroph. Reddish stems (may also be white or coral pink) are 8"–15" tall. Scalelike leaves lack chlorophyll. Yellowish-brown, ½"- to 1"-long flowers are borne in clusters of 2–11. Flowers hang slightly downward when first appearing; as stamens and styles mature, the flowers become horizontal. Fruit is a capsule.

Bloom season: Summer

Range/habitat: Across most of North America in moist forest humus

Comments: *Monotropa* ("one direction") refers to the flowers facing the same direction. *Hypopitys* is from the Greek *hypo* ("under") and *pitys* ("pine tree"), in reference to the plant's habit of growing below pine or coniferous trees. The plant derives nutrients from parasitizing soil fungi that are associated with the roots of other plants. When the flowers are fully mature, they point downward; as the capsules mature, the flowers point upward. Also known as yellow bird's nest, or Dutchman's pipe for the pipelike appearance of the stem and flowers.

WOODLAND PINEDROPS

Pterospora andromedea
Heather family (Ericaceae)

Description: Myco-heterotroph. One to several reddish-brown stems arise 6"–40" tall. Scalelike leaves are reddish brown and basal. The top of the plant bears a dense cluster of small, urn-shaped flowers that are yellowish brown and hang downward on short stalks. Flowering stalk is hairy and sticky to the touch. Fruit is a round capsule.

Bloom season: Summer

Range/habitat: Western North America, across most of Canada, into northern Mexico, and in parts of the northeastern United States in coniferous and mixed forests

Comments: *Pterospora* ("winged seeds") describes the seeds, which have netlike wings on their edge to catch the air and aid in dispersal. *Andromedea* ("Andromeda") is named for the mythical maiden chained to a rock and rescued by Perseus in Greek mythology. Like other myco-heterotrophs, this plant derives nutrients from fungal associations with its roots and the roots of other plants.

JASON KSEPKA

WILD INDIGO

Baptisia tinctoria
Pea family (Fabaceae)

Description: Perennial, bushy growth, 2'–3' tall. Leaves are divided into 3 grayish-green leaflets that are spatula-shaped and ½"–1" long. Yellowish pea-shaped flowers, about ½" long, are borne in short elongated clusters that are 4"–5" long. Fruit is an inflated seedpod that turns black at maturity.

Bloom season: Late spring to midsummer

Range/habitat: Eastern United States in dry meadows and open woods

Comments: *Baptisia* is from the Greek *baptizein* ("to dye"), indicating that some members of the genus were used as an alternative to true indigo dye. *Tinctoria* ("dye") also refers to the use of these dye plants. The plants contain toxins that may cause eye irritation or skin rashes. Gopherweed (*B. lanceolata*) also grows in the southeastern United States and has lance-shaped leaflets that are up to 5" long. Both species attract bees and butterflies as pollinators.

STEVE R. TURNER

PARTRIDGE PEA

Chamaecrista fasciculata
Pea family (Fabaceae)

Description: Annual, with slender stems 1'–3' tall. Leaves are pinnately compound and have 8–18 pairs of ⅔"-long oblong leaflets. Yellow flowers are arranged in small clusters of 2–6 and arise from leaf axils. The yellow flowers are about 1" wide, have 5 rounded petals, and 10 stamens, of which 6 are red and 4 are yellow. Fruit is a seedpod about 2½" long that splits open at maturity.

Bloom season: Late spring to early fall

Range/habitat: Central and eastern United States in sandy dunes, woods, prairies, grasslands, thickets, and recently burned areas

Comments: *Chamaecrista* is from the Greek *chamae* ("low growth") and *crista* ("crested"). *Fasciculata* ("banded" or "bundle of nerves") refers to the action of the leaves closing when touched. As is the case with many members of the Pea family, small microorganisms associated with the root nodules produce nitrogen compounds necessary for the plant's survival. Bees and butterflies are attracted to the flowers; birds eat the seeds, which are ejected from the pods when mature. The red stamens produce reproductive pollen; the 4 yellow stamens produce food pollen to attract pollinators. Also known as sleepingplant due to the folding leaves "going to sleep" when touched.

CHUCK TAGUE

ROBERT WEBSTER

LOW RATTLEBOX

Crotalaria pumila
Pea family (Fabaceae)

Description: Annual or perennial, often low growing, 3"–6" tall, with creeping stems, but may reach 1½' tall. Leaves are alternate pinnately compound, about 1½" long, with 3 inversely lance-shaped to elliptical leaflets that are about the same size. Leaves have small stipules at the base of the leaf stem. Pea-shaped flowers are borne in small clusters of 1–5; flowers have 5 yellow to yellow-orange petals that are streaked with red and have an upper enlarged petal (the banner) and 2 lower ones that are fused together to form the keel. Fruit is an inflated hairy legume about ½" long.

Bloom season: Late summer to mid-autumn

Range/habitat: Somewhat disjointed; grows in the southwestern United States, parts of the southeastern United States, and northern and central Mexico in sandy and disturbed sites, fields, and woodlands

Comments: *Crotalaria* is from a Greek word meaning "castanet" and is the same root for *Crotalus*, the genus of rattlesnakes. *Pumila* ("low growing") refers to the plant's stature. The common name refers to the plant height and the sound the seedpods make when the loose seeds rattle about inside the pods. These plants contain toxic pyrrolizidine alkaloids that are poisonous to humans and livestock.

GOLDEN PRAIRIE CLOVER

Dalea aurea
Pea family (Fabaceae)

Description: Perennial, 1'–3' tall; stems have silky hairs. Pinnately compound leaves are ½"–1½" long, alternate, and with 3–9 linear to egg-shaped leaflets that are ⅜"–¾" long. Leaves are hairy and have pointed tips. Yellow pea-shaped flowers are arranged in cone-like clusters, 1½"–2½" long, with 5-lobed hairy calyces and bright yellow, ¼"-long petals. Fruit is a small, 1-seeded pod.

Bloom season: Mid-spring to early summer

Range/habitat: Central United States and northern Mexico in silty or gravelly prairies, meadows, open woods, brushy areas, and disturbed sites

Comments: *Dalea* honors Samuel Dale (1659–1736), an English apothecary. *Aurea* ("golden") refers to the flower color. Native tribes used the plant to treat stomach upsets and diarrhea. Also known as silktop dalea for the silky flower heads.

TOM LEBSACK

INDIAN RUSHPEA

Hoffmannseggia glauca
Pea family (Fabaceae)

Description: Perennial; often sprawling, but up to 12" tall. Twice pinnately compound leaves have smooth or minutely hairy leaflets. Small leaflets are linear, borne in opposite pairs along short stems that branch off other stems, creating a mat-like appearance. Yellow to salmon-colored flowers have orange and red spotting on the widely spreading, spatula-shaped petals. Flowers have small glandular hairs. Fruit is a curved seedpod.

Bloom season: Mid-spring to late summer

Range/habitat: Southwestern United States in dry desert soils, disturbed areas, and gravelly areas

Comments: *Hoffmannseggia* honors Johann Centurius Hoffmann, also known as Count von Hoffmannsegg (1766–1849), a German botanist and entomologist who coauthored a book on the flora of Portugal. *Glauca* ("bluish gray") refers to the color of the leaves. Native tribes harvested and cooked the edible tubers like potatoes.

LAUREN MCLAURIN

YELLOW PUFF

Neptunia lutea
Pea family (Fabaceae)

Description: Perennial, 3"–4" tall; creeping stems up to 5' long are covered with soft spines. Twice-compound leaves are about 3½" long and divided into 8–18 pairs of leaflets; each leaflet has 14–43 smaller leaflets that are up to ⅓" long. Leaflets close when touched (rapid plant movement known as thigmonasty). Flowers are borne in elongated clusters of 30–60 flowers on terminal stalks that are 3½" long, with 5 tiny petals and 10 protruding stamens. Fruit is a flattened seedpod.

Bloom season: Mid-spring to fall

Range/habitat: Southeastern United States, from Texas to Alabama, in sandy soils in grasslands, prairies, woodlands, and disturbed sites

Comments: *Neptunia* honors Neptune, Roman god of the sea, and refers to the coastal distribution of plants in this genus. *Lutea* ("yellow") refers to the color of the flowers. Also known as yellow sensitive briar. Cattle often forage on this plant because of its lack of spines.

LANE CHAFFIN

TEXAS SENNA

Senna roemeriana

Pea family (Fabaceae)

Description: Perennial, 1'–2' tall, with hairy stems. Leaves are long stalked, lance-shaped, and divided into 2 leaflets that are 1½" long and form a V-shape. Yellow, 1"-wide flowers have 5 sepals, 5 petals, and a cluster of 9–10 stamens that resemble a bunch of bananas. Flowers are bisymmetrical; some petals are slightly larger than the others. Fruit is a seedpod.

Bloom season: Mid-spring to early fall

Range/habitat: South-central United States and northeastern Mexico in fields and open woods; often associated with limestone soils

Comments: *Senna* is from the Arabic *sanā*, which refers to plants that were used as laxatives. *Roemeriana* honors Carl Ferdinand von Roemer (1818–1892), a German geologist who studied Texas geology and collected plants in that region from 1845 to 1847; he is often referred to as the father of Texas geology. The leaves or pods were brewed to make a laxative tea; however, the tea may be toxic if consumed in large quantities. Large bees are common pollinators of these flowers.

MOUNTAIN GOLDEN-BANNER

Thermopsis montana

Pea family (Fabaceae)

Description: Perennial; erect stems rise 1'–2' tall. Numerous compound leaves with 3 smooth, broadly egg-shaped leaflets are 2"–4" long. Yellow pea-shaped flowers are borne in a dense, elongated spike of 5–50 flowers, with each flower about 1" long. Sepals are covered with dense hairs. Fruit is a straight, upright seedpod that is slightly hairy and bluish green.

Bloom season: Late spring to midsummer

Range/habitat: Western United States in moist meadows, disturbed sites, and open woodlands

Comments: *Thermopsis* is from the Greek *thermos* ("lupine") and *opsis* ("similar"), meaning "resembling a lupine," and refers to the flowers resembling those of lupines. *Montana* ("of the mountains") refers to the distribution of these plants. This plant is unpalatable to livestock and therefore may grow in dense clusters. The plants may spread from underground roots and thus cover large patches of ground.

STEVE R. TURNER

GOLDEN CORYDALIS

Corydalis aurea
Fumitory family (Fumariaceae)

Description: Annual or biennial, 2½"–15" tall; often clump forming. Blue-green leaves are compound, up to 3" long, and are several times divided into linear to oblong leaflets, creating a feathery appearance. Flower stems bear few to many (20) yellowish flowers that are ½"–¾" long and have 4 petals in 2 pairs, with 2 small, triangular sepals. Upper and lower petals are unlike in shape—the upper petal has a short, sack-like spur; the lower petal spreads outward and is tongue-shaped. The 2 inner petals are joined together at the tip to form a hood over the style and stamens. Fruit is a cylindrical capsule, about ¾" long, that curls up at maturity.

Bloom season: Early summer to early fall

Range/habitat: Across much of North America in desert grasslands, prairies, gravelly soils, meadows, shrublands, and woodlands

Comments: *Corydalis* ("crested lark") refers to the spur-shaped petal, which resembles the claw of a lark. *Aurea* ("golden") refers to the color of the flowers, which start out erect then droop with age. Also known as golden smoke or scrambled eggs for the flower color.

JIM FOWLER

PALE CORYDALIS

Corydalis sempervirens
Fumitory family (Fumariaceae)

Description: Biennial, 1'–2' tall. Compound leaves are divided into 3–7 segments, up to ¾" long; leaflets are cleft into 2–3 segments, which are again divided into narrow lobes that have rounded tips. Upper leaves are stalkless; the lower leaves have stems. Sack-like flowers are borne at the ends of the flowering stalks and are pink with yellow ends. The pair of sepals are teardrop-shaped. Flowers have 2 pairs of petals—one set is pale pink with yellow lobes; the other is thin and yellow and fits within the fold of the upper pair. Fruit is a long, narrow seedpod.

Bloom season: Late spring to early fall

Range/habitat: Across Canada and the northeastern United States, south to northern Georgia, along rocky or sandy lakeshores and in river basins, disturbed or burned areas, and open woods

Comments: *Corydalis* ("crested lark") refers to the spur-shaped petal, which resembles the claw of a lark. *Sempervirens* is from *semper* ("always") and *virens* ("flourishing or green"), referring to the leaves. Pollination is through ants, other insects, and the wind, although bees and butterflies also visit the flowers and may contribute to pollination. The roots are toxic. Yellow harlequin (*C. flavula*) also grows in the region and has ½"-long yellow flowers with a short spur.

KATIE BYERLY

GIANT ST. JOHN'S WORT

Hypericum ascyron
St. John's Wort family (Hypericaceae)

Description: Perennial, 3'–5' tall. Light green, square stems are unbranched except for the upper one-third of the plant. Lance- to egg-shaped leaves are opposite, up to 4"–5" long, with entire margins. The 5-petaled yellow flowers are borne in groups of 1–5. Each flower is 2"–2½" wide and supports numerous stamens with orange tips and 4–5 styles; petals may be somewhat floppy or slightly twisted. Fruit is a capsule.

Bloom season: Early to late summer

Range/habitat: Southern to northeastern United States in moist stream banks and woods

Comments: *Hypericum* is from the Greek *hyper* ("above") and *eikon* ("picture"), referring to the practice of hanging bunches of flowers of this genus above windows or mounted pictures. *Ascyron* is the name of another *Hypericum* species. Bumblebees are primary pollinators of the flowers, but a host of insects visit the flowers. Some members of this genus were harvested and burned on the eve of St. John's Day, June 24, to ward off evil spirits.

LAURA CLARK

NITS AND LICE

Hypericum drummondii
St. John's Wort family (Hypericaceae)

Description: Annual, with slender stems up to 12" tall; lower stems are reddish brown. Small, needle-like leaves are opposite, ¾" long, and covered with black to yellowish-brown resinous dots. Flowers arise from leaf axils, singly or in small clusters, and are ⅓" wide with 5 orange-yellow petals. Fruit is a seed.

Bloom season: Summer

Range/habitat: Eastern and southern United States in fields, prairies, pastures, open forests, and rocky bluffs

Comments: *Hypericum* is from the Greek *hyper* ("above") and *eikon* ("picture"), referring to the practice of hanging bunches of flowers of this genus above windows or mounted pictures. *Drummondii* honors Thomas Drummond (1790–1835), a Scottish naturalist who collected birds and plants in the southeastern United States. The resinous spots on the leaves give this plant its common name. Common St. John's wort (*H. perforatum*) is an introduced weed from Europe now widely distributed in North America.

STEVE R. TURNER

YELLOW STAR GRASS

Hypoxis hirsuta
Star-Grass family (Hypoxidaceae)

Description: Perennial, low growing, 3"–8" tall. Hairy, grasslike leaves arise from the base and are up to 12" long. Linear leaves have smooth edges and are often taller than the flowering stalks. Thin flowering stalks, which may be upright or recurving, bear ¾"-wide, yellow star-shaped flowers. Six lance- to egg-shaped tepals make up the flower; the 3 outer ones are hairy underneath. The 6 stamens bear spear-shaped anthers. Fruit is a capsule with round black seeds.

Bloom season: Mid-spring to midsummer

Range/habitat: Central and eastern North America in prairies, savannas, open woodlands, forest edges, fens, and disturbed areas

Comments: *Hypoxis* means "a little sour" and was a name Linnaeus gave to this genus of plants, borrowing the name from the close-resembling *Gagea* genus; the name refers to the taste of the leaves. *Hirsuta* ("hairy") refers to the hairs on the leaves, stem, and flowers. Small bees, flies, and beetles are common pollinators on these dainty plants. Formerly in the Lily family (Liliaceae), this plant has been assigned to its own family. The grasslike leaves and yellow star-shaped flowers give this plant its common name. Some tribes brewed an herbal tea from the leaves to treat heart symptoms.

STEVE R. TURNER

SPOTTED BEE BALM

Monarda punctata
Mint family (Lamiaceae)

Description: Perennial, 6"–3' tall, with square stems. Oblong leaves are up to 3" long, toothed along the margins, slightly hairy, pointed at both ends, and aromatic. Flowers are arranged in stacked whorls along an elongated stalk at the end of a stem or from one that arises from the leaf axils. The yellow tubular-shaped flowers are 2-lipped with purple spotting and ¾"–1" long. Subtending the flowers are white leaflike bracts that have a purple tinge. Fruit is a seed.

Bloom season: Early to midsummer

Range/habitat: Eastern North America in dry sites in prairies and coastal plains

Comments: *Monarda* honor Nicolás Bautista Monardes (1493–1588), a Spanish physician and botanist from Seville. *Punctata* ("spotted") refers to the marks on the flowers and bracts. The plants contain thymol, a natural phenol that has antiseptic or antifungal properties and was used medicinally to treat colds, diarrhea, and kidney ailments. Many Monarda species have a single flower cluster per stalk, but spotted bee balm has multiple clusters stacked along a single stem. Wasps, butterflies, and bees are common pollinators attracted to these flowers.

ELEANOR DIETRICH

YELLOW-FLOWERED BUTTERWORT

Pinguicula lutea

Bladderwort family (Lentibulariaceae)

Description: Carnivorous plant; flowering stems may reach 20" tall. Basal rosette of yellowish-green leaves that are oblong to egg-shaped, 2"–3" long, and curl inward longitudinally. Leaves also have sticky surfaces that trap small insects. Bright yellow flowers are irregular, with 5 sepals and 5 petals with notches in the tip. One of the petals forms a spur that projects behind the ½"- to 1⅛"-long flower. Stems and sepals are hairy. Fruit is a rough capsule.

Bloom season: Late winter to late spring

Range/habitat: Southeastern United States in moist or dry soils in bogs, marshes, and savannas—soils that are generally nutrient-poor

Comments: *Pinguicula* is from the Latin *pinguis* ("fat") and refers to the greasy texture of the leaves. *Lutea* ("yellow") refers to the flower color. The sticky substance on the leaves comes from a gland (the peduncular gland) located along the stem; small insects mistake the substance for water droplets. When insects become trapped, the leaves roll inward somewhat, and digestive enzymes are secreted to consume the insect and obtain nitrogen from the plant's prey.

NOLAN EXE

COMMON BLADDERWORT

Utricularia macrorhiza

Bladderwort family (Lentibulariaceae)

Description: Perennial, aquatic carnivorous plant; lacks roots and is free floating or submerged. Alternate, branch-like leaves are ¾"–2" long, 2-parted at the base, and divided several times to form narrow, forked segments; leaves may be underwater but also may grow above the waterline. The leaf's main stem appears zigzagged. Large red bladders arise from the leaf segments. There are 6–20 yellow flowers, which are borne on a stout stem above the water. The snapdragon-like flowers are globe-shaped and have fine red veins. Fruit is a round capsule.

Bloom season: Summer

Range/habitat: Widespread in North America in ponds and slowly moving streams and rivers

Comments: *Utricularia* ("little bladder" or "little bottle") refers to the inflated sacs, which trap insects. *Macrorhiza* ("large roots or root stock") refers to the underwater roots. The bladders on the stems are triggered by small organisms brushing against them. The bladder then opens, and a rush of water carries the organisms into the bladder. The bladder closes, and digestive enzymes go to work on the organisms; the process is quicker than the blink of an eye. Other microorganisms, including bacteria, algae, and diatoms, may live symbiotically in the bladders. Lesser bladderwort (*U. minor*) is another bladderwort that occurs in the northern half of common bladderwort's range.

JIM FOWLER

BLUE-BEAD LILY

Clintonia borealis
Lily family (Liliaceae)

Description: Perennial, 5"–10" tall. Plants may spread by rhizomes and blanket an area. Oval to strap-like leaves (3–5) may be up to 12" long (generally 6"–8") and have parallel venation. Bell-shaped flowers are borne in loose flat-topped clusters and have 6 yellow tepals and 6 stamens. The drooping flowers are about ¾" wide. Fruit is a blue berry, although a white form also exists.

Bloom season: Late spring to midsummer

Range/habitat: Eastern North America from Canada to North Carolina in coniferous, deciduous, or mixed forests and in moist or dry sites such as bogs, swamps, or rocky slopes

Comments: *Clintonia* honors DeWitt Clinton (1769–1828), former governor of New York. *Borealis* ("of the north") refers to its distribution in boreal forests. The colorful fruits are poor tasting and semi-toxic. The roots contain diosgenin, a steroid compound used in estrogen therapy. Also known as yellow Clintonia or yellow beadlily, to name but a few of its common names.

STEVE R. TURNER

YELLOW TROUT LILY

Erythronium americanum
Lily family (Liliaceae)

Description: Perennial, 3"–6" tall. Two glossy, tongue- or strap-like basal leaves about 6" long arise from the base of the flowering stalk and may be mottled with brown. The single nodding or bell-shaped flower is borne on a naked stem and has 6 yellow tepals that curl partially backward and may show some purple on the undersides. Fruit is a capsule.

Bloom season: Mid- to late spring

Range/habitat: Eastern North America in moist woods, slopes, and along streams

Comments: *Erythronium* is from the Greek *erythros* ("red") and refers to a red dye made from another species in this genus. *Americanum* ("of North or South America") indicates the distribution of this plant in Canada and the United States. The plants may grow in dense patches; plants with a single leaf will not flower. Also known as yellow dog-tooth violet for the shape of the leaves.

MATT BERGER

GLACIER LILY

Erythronium grandiflorum
Lily family (Liliaceae)

Description: Perennial, from bulb 1½"–2" wide; plants are 6"–15" tall. Large, non-mottled basal leaves are arranged in pairs and clasp the flowering stem's base. Atop the leafless, flowering stem is a single (sometimes a pair) golden-yellow flower that hangs downward. The 6 tepals curve upward, while the large yellow stamens protrude downward. The 1"-long, club-shaped capsule contains papery seeds.

Bloom season: Late spring to midsummer

Range/habitat: Western North America in subalpine meadows, openings, and along forest edges

Comments: *Erythronium* is from the Greek *erythros* ("red") and refers to a red dye made from another species in this genus. *Grandiflorum* ("large-flowered") refers to the flower's size. Plants may bloom near the edges of snowfields or glaciers; hence the common name. The plants are able to photosynthesize under the snow, and the leaves and buds often push up through snow. Also known as yellow avalanche lily or yellow fawn lily, for the large mottled leaves that resemble the ears of a fawn.

YELLOW BELLS

Fritillaria pudica
Lily family (Liliaceae)

Description: Perennial; grows 4"–12" high from a small white bulb. Stem bears 2–6 narrow, strap-like leaves, ½"–6" long. Flowering stalk bears 1–3 yellow, bell-shaped flowers that hang downward. Flowers have 6 tepals. Fruit is a round or egg-shaped capsule.

Bloom season: Early to late spring

Range/habitat: Western North America in dry grasslands, woodlands, and open meadows

Comments: *Fritillaria* is from the Latin *fritillus* ("a dice box") and refers to the shape of the capsules. *Pudica* ("bashful") refers to the pendulous flowers. Lewis and Clark, during their Corps of Discovery Expedition, noted that the local Northwest tribes collected and ate the onion-like bulbs. Bees, flies, and beetles pollinate the early-season flowers.

ELEANOR RAY

CANADA LILY

Lilium canadense
Lily family (Liliaceae)

Description: Perennial, 3'–8' tall. Narrowly oval to lance-shaped leaves (3–8), up to 6" long, are borne in whorls along the stem; there may be a few alternate leaves as well along the stem. Flowering stalks bear 1–5, rarely up to 20, flowers that range from yellow to orange-red with dark spots. Nodding, trumpet-shaped flowers are about 2½" long; tepals are curved backward but don't touch the tube. Fruit is a capsule about 2" long.

Bloom season: Early to midsummer

Range/habitat: Eastern North America in moist meadows and woodland edges

Comments: *Lilium* is the Latin name for this genus. *Canadense* ("of Canada") refers to the distribution of the plant. Native Americans harvested the roots and buds for food. Michigan lily (*L. michiganense*) has tepals that curve backward and touch the flowering tube. Large butterflies are common pollinators of these flowers. Also known as meadow lily or wide yellow lily.

JUDY PERKINS

SAND BLAZING STEM

Mentzelia involucrata
Blazing Star family (Loasaceae)

Description: Annual; stems 1½"–12½" tall. Stems and leaves are covered with rough hairs. Larger leaves are mostly basal, ¾"–7" long, elliptical to lance-shaped, and with irregularly toothed or lobed margins. Creamy yellow bell-shaped flowers are borne singly and are 2½" wide, with 5 sepals and 5 overlapping petals. Four to 5 egg-shaped, white to dark green bracts with 3–10 different size lobes subtend the flowers. Fruit is a capsule with ash-white seeds.

Bloom season: Midwinter to late spring

Range/habitat: Southwestern United States and Baja California in open rocky or sandy sites, alluvial fans, washes, shrublands, and rocky slopes

Comments: *Mentzelia* honors Christian Mentzel (1622–1701), a German botanist and physician. *Involucrata* ("involucre") refers to the bracts below the flowers. Native peoples harvested the seeds as a food source. In a case of mimicry, sand blazing stem flowers produce nectar to attract *Xeralictus* bees. Similar-looking flowers of ghost flower (*Mohavea confertiflora*) do not produce nectar but have marks on the flowers that resemble female *Xeralictus* bees to attract male bees.

GIANT BLAZING STAR

Mentzelia laevicaulis
Blazing Star family (Loasaceae)

Description: Perennial or biennial; plants up to 3' tall. Stems are many branched and white. Basal leaves are lance-shaped, up to 4" long, toothed along the margins, and rough textured. Upper leaves are smaller. Starlike yellow flowers are 2"–3" wide, with 5 pointed petals and a short, green bract between the petals. Numerous thread-like stamens project above the flower. Fruit is a capsule with tiny black seeds.

Bloom season: Summer through early fall

Range/habitat: Western North America in sandy or gravelly sites

Comments: *Mentzelia* honors Christian Mentzel (1622–1701), a German botanist and physician. *Laevicaulis* ("smooth stem") describes the feel of the stem, which is in contrast to the rough leaves. The flowers first open in the evening and close in the morning until the petals mature; afterward, the flowers remain open during the day. Flies, beetles, moths, small bees, and wasps may be observed wandering through the flower's forest of stamens in search of pollen.

ELEANOR DIETRICH

GOLDEN COLICROOT

Aletris aurea
Bog Asphodel family (Nartheciaceae)

Description: Perennial, 10"–36" tall. Elliptical to lance-shaped basal leaves are 4"–8" long. Yellow, barrel- or bell-shaped flowers are ¼" long, arranged along the flowering stalk, and have 6 lobes. The flower's texture is mealy. Fruit is a capsule.

Bloom season: Mid-spring to late summer

Range/habitat: Southeastern United States in savannas, bogs, and pine woodlands

Comments: *Aletris* is from the Greek for a "woman who grinds," mainly a slave woman who ground grain or corn into a mealy texture. *Aurea* ("golden yellow") refers to the color of the flowers. The common name refers to the use of the roots to treat colic.

STEVE R. TURNER

AMERICAN LOTUS

Nelumbo lutea

Lotus family (Nelumbonaceae)

Description: Perennial; plants range 2½'–5' tall (underwater). Round leaf blades are 8"–17" wide. A stout petiole elevates the blades above the water, connecting to the center of the leaf like a parasol; these petioles may be up to 6½' long. Light yellow cup-shaped flowers are 6"–12" wide, have 22–25 petals, and have a stout, inversely conical-shaped receptacle with cavities that contain pistils. Fruit is a flat-topped structure with hard round seeds about ½" wide embedded in the woody receptacle.

Bloom season: Late spring to late summer

Range/habitat: Eastern North America, central United States, and California, south in Central and South America and the West Indies in lakes, ponds, canals, and areas prone to flooding

Comments: *Nelumbo* is a Sinhalese name meaning "holy lotus" and refers to the sacred lotus (*N. nucifera*) of Asia and Australia. *Lutea* ("yellow") describes the flower color. The plant is a larval host for the American lotus borer (*Ostrinia penitalis*), a species of moth. The larvae feed on the leaves under a silken net then bore into the leaf's petiole, where they pupate. Native Americans expanded the range of this plant as they carried seeds or plants with them as a food resource. The edible seeds are often called "alligator corn"; the leaves and roots were also eaten. When mature, the receptacle breaks off and floats in the water, dispersing the seeds.

YELLOW SAND VERBENA

Abronia latifolia

Verbena family (Nyctaginaceae)

Description: Low-growing perennial with prostrate stems up to 3' long. Stems have sticky hairs. Leaf blades are round to kidney- or oval-shaped and are oppositely arranged along the stems. Flower heads contain several to many tubular, ½"-long yellow flowers that flare open at the mouth. Fruits are cylindrical seeds adorned with thick, keel-like projections.

Bloom season: Mid- to late summer

Range/habitat: Western North America in coastal dunes and sandy beaches

Comments: *Abronia* is from the Greek *abros* ("delicate" or "graceful"), which describes the bracts below the fragrant, slim flowers. *Latifolia* ("broad leaves") refers to the leaf shape. Sand grains adhere to the sticky hairs on the stems and leaves; hence the common name. Shifting sands may bury portions of the plant, sometimes giving the appearance that the flowers are sprouting from the sand. Pink sand verbena (*A. umbellata*) has pinkish flowers and grows in western North America.

YELLOW POND LILY

Nuphar polysepla
Water Lily family (Nymphaeaceae)

Description: Aquatic perennial; egg- to heart-shaped leaf blades are 6"–22" long and float on the water surface. Yellow flowers are 2" across and composed of inconspicuous petals and small green, outer sepals. Larger, inner sepals are yellow but may have a purple or greenish tinge. Center of the flower is a large knob-like, long-stalked stigma. Fruit is an oval, ribbed capsule that releases seeds in a jellylike mass when mature.

Bloom season: Late spring through summer

Range/habitat: Western North America in ponds, lakes, or other bodies of standing water

Comments: *Nuphar* is from the Arabic word *naufar*, a name for a water lily. *Polysepla* ("many sepals") refers to the flowers. Many Northwest tribes used the root for medicinal purposes and ate the seeds. Caddis flies and beetles are attracted to the flower's nectar as pollinators; the plants are a preferred larval host for the water lily leaf beetle (*Galerucella nymphaea*). Also known as wokas or wocus.

LAVENDER-LEAF SUNDROPS

Calylophus lavandulifolius
Evening Primrose family (Onagraceae)

Description: Perennial; low growing, up to 9" tall. Linear to inversely lance-shaped leaves are ¼"–1½" long and covered with dense hairs. Leaf bases are wedge-shaped. Solitary yellow flowers arise in the leaf axils; the corolla tube is ¾"–3" long, and the 4 flaring petals have ruffled edges. Fruit is a capsule.

Bloom season: Late spring to late summer

Range/habitat: Central and southwestern United States in desert shrublands and open woodlands

Comments: *Calylophus* is from the Greek *caly* ("calyx") and *lophus* ("a crest"), referring to the small projection on the sepals. *Lavandulifolius* ("lavender-like leaves") refers to the resemblance of these leaves to those of lavender. The flowers bloom during the day and attract butterflies and bees as pollinators. Toothed or yellow sundrops (*C. serrulatus*) grows in the central United States and Canada and has sharply toothed leaves.

STEVE R. TURNER

SEEDBOX

Ludwigia alternifolia
Evening Primrose family (Onagraceae)

Description: Perennial, 2'–4' tall, with multiple branches. Alternate lance-shaped leaves are up to 4" long and broadest in the middle. Yellow flowers are borne singly on short stalks from leaf axils; the ¾"-wide flower has 4 green, red-tinged triangular sepals, 4 yellow petals, and 4 stamens. The petals drop off soon after flowering, sometimes the same day the flowers open. Fruit is a squarish, ¼"-wide capsule.

Bloom season: Summer

Range/habitat: Eastern North America in wet areas along streams, swamps, marshes, prairies, swamps, ditches, and roadsides

Comments: *Ludwigia* honors Christian Ludwig (1709–1773), a professor of botany in Leipzig, Germany. *Alternifolia* ("alternate leaves") refers to the arrangement of the leaves. The common name is for the boxlike capsules, which sound like small rattles when the seeds mature; another common name is rattlebox.

DAVID LEGROS

EVENING PRIMROSE

Oenothera biennis
Evening Primrose family (Onagraceae)

Description: Biennial, up to 7' tall, with a stout stem that is tinged with red and sometimes hairy. Basal leaves are lance-shaped, 4"–12" long, and irregularly toothed along the margins. Stem leaves are shorter, up to 6" long, lance-shaped, and hairy. Yellow, disk-shaped flowers arise along a long stalk, are about 2½" wide, and have 4 notched petals and 8 stamens. Fruit is a long, slender capsule.

Bloom season: Midsummer to early fall

Range/habitat: Occurs across much of North America except for some of the Rocky Mountain states in prairies, fields, pastures, disturbed places, and along stream banks

Comments: *Oenothera* ("wine scented") refers to some species in this genus being used to make wine. *Biennis* ("biennial") describes the plant's life cycle. Sphynx moths are attracted to the flowers, which open at night and give off a lemon scent. Decoctions were made from the leaves and flowers to treat respiratory ailments. The long-lived seeds have been known to germinate after 70 years.

TANSY-LEAF EVENING PRIMROSE

Taraxia tanacetifolia
Evening Primrose family (Onagraceae)

Description: Perennial, low growing; lacks a stem. Long leaves are highly lobed or divided; leaves are up to 4" long. Yellow flowers are 1"–2" wide, arise from a basal cluster of leaves, and have 4 petals. Petals form a long tube, and the 4 sepals become reflexed as the flower matures. Fruit is a leathery capsule.

Bloom season: Late spring to late summer

Range/habitat: Western United States in grasslands, shrublands, meadows, sandy sites, and open woodlands

Comments: The derivation of *Taraxia* is thought to be from the Greek *taraxia* ("eye disorder") and may have some relationship to the dandelion genus, *Taraxacum*. *Tanacetifolia* ("tansy-leaved") refers to the tansy-like leaves. Diffuseflower evening primrose (*T. subacaulis*) also grows in the western United States but has broadly lance-shaped to oval leaves.

STEVE R. TURNER

YELLOW LADY'S SLIPPER

Cypripedium parviflorum
Orchid family (Orchidaceae)

Description: Perennial, stems up to 30" tall; often clump forming. Stems and leaves are hairy. Plants bear 3–6 oval-shaped leaves with parallel venation, up to 9" long and pointed at the tip. One to 2 flowers are borne on a stem; the yellow flowers are 1¼"–2¼" long with an inflated slipper-like petal surrounded by 2 twisted, narrow petals that are 2"–3½" long. Two broad yellowish-green to greenish-brown sepals sit above and below the slipper. Fruit is a capsule.

Bloom season: Mid-spring to midsummer

Range/habitat: Widespread across eastern and northern North America, the Rocky Mountain states, and in isolated areas in the southwestern United States in prairies, wetlands, meadows, woodlands, and along stream banks

Comments: *Cypripedium* is from the Greek *Kypris* ("a name for Aphrodite") and *pedilon* ("slipper"), which translates to "Aphrodite's slipper." *Parviflorum* ("small flower") refers to the size of the beautiful flowers. Modern-day herbalists use cultivated plants as a sedative or pain reliever.

JIM FOWLER

TUBERCLED ORCHID

Platanthera flava
Orchid family (Orchidaceae)

Description: Perennial, up to 2½' tall. The 2–3 elliptical to lance-shaped leaves are 3"–8" long, narrow, and with entire margins. An elongated flowering stalk bears numerous, up to 60, greenish-yellow flowers that are ¼" wide. The center of the flower's lower petal has a prominent tubercle at its base, as well as 2 small lobes. A club-shaped spur projects outward from the back of the flower. Fruit is a capsule.

Bloom season: Mid-spring to mid-fall

Range/habitat: Central and eastern North America in wet meadows, wooded swamps, thickets, and woodlands

Comments: *Platanthera* ("broad anther") describes the shape of the anther. *Flava* ("yellow") refers to the flower color. Mosquitoes in the *Aedes* genus and small moths are the primary pollinators of these flowers. The tiny seeds are wind dispersed.

STEVE R. TURNER

YELLOW OWL-CLOVER

Orthocarpus luteus
Broomrape family (Orobanchaceae)

Description: Annual, grows up to 16" tall; dense glandular hairy stems are yellow-green or purple. Linear leaves are alternate, about 2" long, stiff, and pointing upward; upper leaves may be divided into 3 lobes. Flowering stalk bears a dense cluster of yellow flowers that have large, 3-lobed green bracts from which the flowers emerge. Yellow flowers are club-shaped and 2-lipped. The small upper lip curves over the top; the lower lip forms a 3-lobed pouch. Flowers are ½" long; petals have small dots on the outside. Fruit is a small capsule.

Bloom season: Mid- to late summer

Range/habitat: Much of western and central North America in prairies, meadows, sagebrush steppe, and woodlands

Comments: *Orthocarpus* is from the Latin *orth* ("straight") and *karpos* ("fruit"), referring to the symmetrical capsules. *Luteus* ("yellow") refers to the flower color. The common name refers to small eyelike spots on the petals or the roundish, head-like corolla with the eye spots resembling an owl peering between several branches. These annuals are semiparasitic on neighboring plants.

CHUCK TAGUE

BRACTED LOUSEWORT

Pedicularis bracteosa

Broomrape family (Orobanchaceae)

Description: Perennial, often 2'–3' tall. Divided, fernlike leaves are 1"–5" long and have saw-toothed edges. Flowers arise in a dense elongated cluster at the end of the flowering stalk. Individual flowers are yellowish to brownish red or purple and have an upper lip that forms a hood and a leaf-like bract at their base. The hood may or may not have a short beak. Fruit is a curved capsule.

Bloom season: Summer

Range/habitat: Western North America in meadows, thickets, forest edges, or open forests

Comments: *Pedicularis* ("relating to lice") refers to the old belief that grazing livestock became plagued with lice after foraging in fields infested with a European species of *Pedicularis. Bracteosa* ("with well-developed bracts") refers to the flower's leaflike bracts. Also known as wood betony or fernleaf, for the shape of the leaves.

COMMON YELLOW WOOD SORREL

Oxalis stricta

Wood Sorrel family (Oxalidaceae)

Description: Annual or perennial, 3"–12" tall. Stems have fine hairs appressed to the surface. Alternate leaves are divided into 3 rounded to heart-shaped leaflets that open and close flat, depending on the sun; leaflets are ¾"–1¼" wide. Yellow flowers have 5 sepals, 5 petals, 10 stamens, are ¼"–½" wide, and are borne in small clusters (2–7) on a stalk about 3" long. Fruit is a banana-shaped seedpod that tosses the seeds several feet away when it opens.

Bloom season: Late spring to early fall

Range/habitat: Central and eastern United States in fields, lawns, woodlands, and disturbed areas

Comments: *Oxalis* is from the Greek *oxys* ("sour") and refers to the edible, although sour-tasting leaves. *Stricta* ("upright") refers to the growth of the stems, which may recline when mature. The opening and closing of the leaves is called nyctinasty and is common in the genus. Southern wood sorrel (*O. dillenii*) is very similar but has 2 flowers per cluster and white ridges on the seeds.

KAREN E. ORSO

BROWN'S PEONY

Paeonia brownii

Peony family (Paeoniaceae)

Description: Perennial, 8"–24" tall. Fleshy leaves are compound once or twice, and the leaflets are deeply lobed. Heavy blossoms, 3"–4" wide, weight the flowering stems down, often to the ground. Five to 6 oval, green or reddish sepals surround the 5–10 white to maroon petals that are edged with yellow. Numerous stamens surround the styles that elongate into fruits. Fruit is a 2"- to 3"-long seedpod.

Bloom season: Mid-spring to early summer

Range/habitat: Western United States in sagebrush plains, open pine woodlands, and meadow edges

Comments: *Paeonia* is from Paeon, the Greek physician to the gods. *Brownii* honors Robert Brown (1773–1858), a British botanist who led expeditions to collect plants in Australia in the mid-1850s. These peonies produce abundant pollen but little nectar, yet they still attract bees, wasps, and ants as pollinators.

CHUCK TAGUE

MEXICAN PRICKLY POPPY

Argemone mexicana

Poppy family (Papaveraceae)

Description: Annual, 8"–18" tall; smooth or prickly stems exude a bright yellow sap when cut. Greenish-white leaves are deeply lobed; the lobes have prickly spines, are alternate, and up to 7¾" long. Cup-shaped flowers are bright yellow, about 2½" wide, and have 6 petals and numerous stamens; sepals are spine-tipped. Fruit is a spine-covered seedpod.

Bloom season: Mid-spring to early summer

Range/habitat: Southeastern United States to South America and the Caribbean in dry soils, meadows, fields, and disturbed places. Plants have expanded their range and naturalized in many states.

Comments: *Argemone* is from the Greek *argema* ("cataract"), a disorder of the eyes that a related species was used to treat. *Mexicana* ("of Mexico") indicates the plant's distribution. When cut, the seedpods ooze a yellow sap. The seeds contain toxic alkaloids, although when the Spanish arrived in the New World, they used the seeds as a laxative.

JIM FOWLER

WOOD POPPY

Stylophorum diphyllum
Poppy family (Papaveraceae)

Description: Perennial, 1'–1½' tall. Stems exude a yellow sap when crushed. Opposite, pinnately compound leaves are hairy below, deeply lobed, and many-toothed along the margins. Yellow flowers, 1½"–2" wide, have 4 petals, 2 sepals that drop off early, numerous stamens, and a single, stout and knobby style. Flowers arise from a pair of leaves at the top of the stem. Fruit is a hairy seedpod.

Bloom season: Early spring to early summer

Range/habitat: Eastern North America along stream banks and in moist woodlands

Comments: *Stylophorum* is from the Greek *stylos* ("style") and *phorus* ("bearing"), in reference to the long, columnar style. *Diphyllum* ("two leaves") refers to the upper pair of leaves. Native Americans used the yellow sap as a dye. Also known as celandine poppy.

STEVE R. TURNER

YELLOW PASSION-FLOWER

Passiflora lutea
Passion-flower family (Passifloraceae)

Description: Perennial vine; stems up to 15' long. Trilobed leaves are 1'–2½' long with shallow rounded lobes and entire margins and lack hairs. Greenish-yellow or whitish flowers are about 1" wide and have 5 stamens, a central pistil, and long, slender petals. Fruit is a black or purple berry.

Bloom season: Late spring to fall

Range/habitat: Eastern and south-central United States in thickets, wetland edges, and woodland forests

Comments: *Passiflora* means "passion flower." When Spanish explorers observed this plant in the New World, the flowers reminded them of the crown of thorns of the crucifixion or Passion of Christ. *Lutea* ("yellow") refers to the color of the flowers. The passionflower bee (*Anthemurgus passiflorae*) feeds exclusively on the pollen from this plant; several butterflies also use this host plant for their larvae, including the gulf fritillary (*Agraulis vanillae*).

JOHN POLITES

YELLOW MONKEY-FLOWER

Erythranthe guttata

Monkey-flower family (Phrymaceae)

Description: Annual or perennial; plant varies in size from 2" to 24" tall. Oval-shaped leaves grow in pairs. Leaves may have hairs or smooth surfaces and may have irregularly lobed or toothed margins. Lower leaves are stalked; upper leaves clasp the stem. Borne on long stalks, trumpet-shaped flowers are ¼"–2" long and 2-lipped, with 1 large or several smaller crimson spots on the lower lip. Fruit is a capsule.

Bloom season: Late spring and summer

Range/habitat: Western North America in moist sites, seeps, meadows, riverbanks, and coastal splash zones

Comments: *Erythranthe* is from the Latin word for "foreign" or "foreigner." *Guttata* ("spotted") refers to the spots on the flower's lower lip. When an insect such as a bee enters the flower, it contacts the stigma. The 2 lobes then fold together and press against the roof of the flower. This forces the insect to contact the anthers, which dust the insect with pollen as it searches for nectar. The lower lip's spots act as nectar guides, directing pollinators into the flower.

STEVE R. TURNER

CLAMMY-HEDGE-HYSSOP

Gratiola neglecta

Plantain family (Plantaginaceae)

Description: Annual, 3"–12" tall, with hairy stems. Lance- to spatula-shaped opposite leaves are ¾"–2" long, covered with fine hairs, and with several teeth along the margins; lower leaves are narrower. A single tubular flower, white at the tip and greenish yellow below, rising on a short stem from a leaf axil, is about ⅓" long with 4–5 lobes. The lower 3 corolla lobes are notched, and the upper lobes may be fused as one and larger than the lower 3, or be notched into 2 lobes that are shorter than the lower lobes; the upper lobes have fine yellow hairs. There are 2 fertile and 2 sterile stamens. Fruit is an oval capsule.

Bloom season: Late spring to early fall

Range/habitat: Widespread throughout the United States and southern Canada along stream banks, wet meadows, floodplains, disturbed areas, and lake margins

Comments: *Gratiola* is from the Latin *gratia* ("agreeable" or "loveliness") in reference to the medicinal qualities of certain species in this genus. *Neglecta* ("neglected") refers to the plant's small stature and weedy appearance.

JUDY PERKINS

GHOST FLOWER

Mohavea confertiflora

Plantain family (Plantaginaceae)

Description: Annual, 4"–16" tall, with hairy stems. Lance- to egg-shaped hairy leaves are alternate, somewhat thick, and up to 4" long. Translucent cup-shaped flowers are about 1½" wide, with ragged-edged petals and pink to purplish spots on the insides of the petals and a darker red spot at the base. Two yellow stamens are located over these dots. Fruit is a capsule with numerous small black seeds.

Bloom season: Late winter to mid-spring

Range/habitat: Southwestern United States and northern Mexico in dry sandy or gravelly washes and slopes

Comments: *Mohavea* ("of the Mohave Desert or River") refers to this plant's distribution and where John C. Frémont first collected these plants for science. *Confertiflora* ("with crowded flowers") refers to the densely packed flower cluster. The flowers do not produce nectar but mimic the flowers of another desert flower, *Mentzelia involucrata*, to attract pollinators. The ghostly translucent color of the flowers gives the plant its common name.

ALEX HEYMAN

GOLDEN GILIA

Leptosiphon chrysanthus

Phlox family (Polemoniaceae)

Description: Annual, up to 6" tall, with thin stems that are smooth or hairy. Linear leaves are opposite, smooth or hairy, generally 3–5 per stem, and about ½" long. Funnel- to tubular-shaped flowers are golden yellow, with 5–6 lobes and 6 stamens. Fruit is a seed.

Bloom season: Early spring to early summer

Range/habitat: Southwestern United States in dry desert flats, shrublands, and woodlands

Comments: *Leptosiphon* is from the Greek *lepto* ("thin") and *siphon* ("a tube"), in reference to the corolla. *Chrysanthus* ("golden flowers") refers to the flower color. When conditions are right, these plants may carpet the desert. Also known as golden desert-trumpets for the shape and color of the flowers.

WILLIAM MCFARLAND

CANDYROOT

Polygala nana

Milkwort family (Polygalaceae)

Description: Annual or biennial, up to 6" tall; low growing and clump forming. Spatula-shaped leaves are ¾"–2½" long and form a basal rosette. Lemon-yellow flowers are borne in a dense cluster that is ¾"–1½" long and shaped like a thimble. Fruit is a capsule.

Bloom season: Generally late spring through summer, but may bloom anytime during the year

Range/habitat: Southern United States from Texas to Florida in wet meadows, swales, and pine woodlands

Comments: *Polygala* is from the Greek *polys* ("many") and *gala* ("milk"), referring to the old belief that cows that foraged on the leaves of plants in this genus would produce more milk. *Nana* ("small") refers to the stature of the plant. The roots have a licorice-like flavor; hence the common name. Ants are the primary dispersers of the seeds. Yellow milkwort (*P. lutea*) is similar but taller.

CUSHION BUCKWHEAT

Eriogonum ovalifolium

Buckwheat family (Polygonaceae)

Description: Mound-forming perennial, 2"–16" across, with flowering stems up to 14" tall. Basal leaves are round or spatula-shaped and covered with woolly hairs. Leaf blades are ¾"–2½" long, and petioles are up to 2" long. Leafless flowering stalks bear a rounded cluster of small, yellow, white, or pinkish flowers striped with purple; the 6 petallike segments are similar. Fruit is a seed.

Bloom season: Early spring to midsummer

Range/habitat: Western North America in open areas in shrublands to alpine areas

Comments: *Eriogonum* ("woolly knees") refers to the hairs growing on the stem and leaf joints. *Ovalifolium* ("oval leaves") refers to the shape of the leaves. A highly variable species across its range, this species attracts bees as pollinators. Also known as tufted wild buckwheat.

STEVE R. TURNER

SULPHUR BUCKWHEAT

Eriogonum umbellatum

Buckwheat family (Polygonaceae)

Description: Mound-forming perennial, 3"–3' tall. Basal leaves are egg- to spoon-shaped, arise on slim stalks, and are green above and gray and woolly below. From the basal cluster of leaves arise flowering stems that are 2"–20" tall. Several narrow, leaflike bracts are arranged below the flower clusters. Numerous small, yellowish flowers that may be tinged with pink are clustered into umbrellalike forms. Individual flowers have 6 lobes and stamens that extend beyond the flower opening. Fruit is a 3-angled seed.

Bloom season: Late spring to late summer

Range/habitat: Western North America in dry, rocky or sandy sites

Comments: *Eriogonum* ("woolly knees") refers to the hairs growing on the stem and leaf joints. *Umbellatum* ("umbrellalike") refers to the shape of the flower cluster. The common name refers to the color of the flowers. This is another highly variable member of the *Eriogonum* genus. The plants are larval hosts for a variety of butterflies, including bramble hairstreak (*Callophrys dumetorum*), lupine blue (*Icaricia lupini*), and Mormon metalmark (*Apodemia mormo*).

WATER SMART-GRASS

Heteranthera dubia

Pickerel-weed family (Pontederiaceae)

Description: Annual or perennial, 1"–2" tall; stems and leaves mostly growing in the water or along the surface or buried in sediments. Flat, ribbonlike leaves are up to 6" long when submerged, but shorter when growing above the water. A 4"-long flowering tube gives rise to a yellow, spiderlike flower that has 6 tepals and is ½"–¾" wide; the flower's center has 3 thick stamens and a style. Fruit is a capsule with winged seeds.

Bloom season: Mid- to late summer

Range/habitat: Over much of North America in sandy or muddy riverbanks and shorelines of lakes or ponds, submersed or on the surface

Comments: *Heteranthera* ("with different anthers") describes the stamens. *Dubia* ("doubtful") refers to the plant's variable growth not conforming to a regular pattern. This aquatic plant may form thick mats of vegetation; plants growing in deeper water produce self-fertilizing flowers. The yellow flowers barely last a day.

STEVE R. TURNER

FRINGED LOOSESTRIFE

Lysimachia ciliata
Primrose family (Primulaceae)

Description: Perennial, 1'–4' tall, with upright or sprawling smooth stems. Lance- to egg-shaped leaves are opposite, up to 6" long and broad, hairless (may have some hairs along the margins), and borne on 1½"-long hairy stems. Star-shaped yellow flowers arise on short stalks from leaf axils, nod downward, and are ½"–1" wide. Flowers have 5 lance-shaped green sepals, 5 yellowish petallike lobes that taper to a tip, 5 stamens, and a slender style. Fruit is a globe-shaped capsule.

Bloom season: Early summer to early fall

Range/habitat: Across much of North America in moist sites in swamps, marshes, thickets, wet prairies, moist woods, and along stream banks

Comments: *Lysimachia* may be derived from the Greek *lys* ("loose" or "a release from") and *mach* ("a fight or strife") and is sometimes attributed to Lysimachus, King of Thrace, who was chased by a bull but calmed it down by waving a sprig of a related plant in the bull's face. *Ciliata* ("small hair" or "eyelash") refers to the hairs on the leaf stem. The flowers produce an oil and pollen that is harvested by the melittid bee (*Macropis steironematis*) for its larvae.

STEVE R. TURNER

YELLOW MARSH MARIGOLD

Caltha palustris
Buttercup family (Ranunculaceae)

Description: Perennial, 8"–24" tall, with hollow stems. Round to heart-shaped leaves are alternate, glossy green, up to 7" long, and have a deep notch at the base; upper leaves are smaller. Cup-shaped flowers are 1"–2" wide, with 5–9 yellow petallike sepals and numerous stamens. Fruit is a seedpod.

Bloom season: Mid-spring to early summer

Range/habitat: Northern Hemisphere temperate regions, from Alaska to Newfoundland and south to the Carolinas in marshes, swamps, fens, wet meadows, and along stream banks

Comments: *Caltha* ("cup of goblet") describes the shape of the flower. *Palustris* ("marsh loving") refers to the plant's preferred habitat. The flower's buds are edible and eaten like capers. Hoverflies (Syrphidae) are common pollinators of these flowers. Also known as cowslip.

JANEL JOHNSON, NEVADA DIVISION OF NATURAL HERITAGE

ESCHSCHOLTZ'S BUTTERCUP

Ranunculus eschscholtzii

Buttercup family (Ranunculaceae)

Description: Perennial, 4"–10" tall, but may be almost prostrate. Basal, round to heart-shaped leaves are divided into segments with round blades; blades may have shallow lobes. Flowering stems bear several 1½"-wide yellow flowers with 5 petals, with a gap between the petals. Fruit is a seed borne in a dense cluster.

Bloom season: Mid- to late summer

Range/habitat: Western North America in moist areas in mountain meadows or along subalpine and alpine streams

Comments: *Ranunculus* is derived from the Latin *rana* ("frog"), for the aquatic or moist habitat preference of the genus. *Eschscholtzii* honors Johann Friedrich von Eschscholtz (1793–1831), a German surgeon and professor who accompanied Otto von Kotzebue on his 1815–1818 and 1823–1826 expeditions to North America.

REUVEN MARTIN

YELLOW WATER BUTTERCUP

Ranunculus flabellaris

Buttercup family (Ranunculaceae)

Description: Aquatic to terrestrial perennial; entire plant is 9"–36" long. Light green stems bear alternate leaves that are palmately lobed and 1"–3½" long and 1"–3" wide. Aquatic plants have deeply divided leaves with linear segments; the overall shape is fan- to kidney-shaped. Emergent or terrestrial plants have lobed leaves with segments that are more oblong-shaped. Yellow flowers are about ¾" wide with 5–6 petals and numerous stamens. Fruits are seeds with straight beaks.

Bloom season: Late spring to midsummer

Range/habitat: Across much of western, northern, and eastern North America (absent in some southwestern and southeastern states) in shallow ponds, swamps, edges of creeks or slow-moving rivers, wetlands, and ditches

Comments: *Ranunculus* is derived from the Latin *rana* ("frog"), for the aquatic or moist habitat preference of the genus. *Flabellaris* is from the Latin *flabellum* ("a small fan") and *aris* ("pertaining to"), referring to the fanlike shape of some of the leaves. The leaves are highly variable and were used by some northern tribes as a snuff for head colds. Also known as large yellow water crowfoot, referring to the flower size and color, aquatic habitat, and shape of the leaves, which often resemble the outline of a crow's foot.

RAVEN TENNYSON

SAGEBRUSH BUTTERCUP

Ranunculus glaberrimus
Buttercup family (Ranunculaceae)

Description: Perennial, low growing, 2"–6" tall. Basal leaves are fleshy and either broad with shallow lobes or divided into deeper lobes. Short flowering stalks bear 1"-wide, bright yellow flowers that have 4–7 petals and numerous stamens and pistils. Each petal has a nectar gland at the base. Fruit is a seed.

Bloom season: Early to late spring

Range/habitat: Western North America in sagebrush steppe or pine woodlands

Comments: *Ranunculus* is derived from the Latin *rana* ("frog"), for the aquatic or moist habitat preference of the genus. *Glaberrimus* ("without hairs") refers to the smooth leaves. Pollinators such as beetles, flies, wasps, and other insects visit the plentiful flowers for either nectar, located at the base of the petals, or pollen, which is produced in abundance by the stamens.

JIM FOWLER

HISPID BUTTERCUP

Ranunculus hispidus
Buttercup family (Ranunculaceae)

Description: Perennial, with stems that trail along the ground; plants are 8"–18" tall. Stems, leaves, and sepals are covered with stiff hairs. Basal leaves are egg-shaped to rhombic and are either 3-parted or separated into 3 leaflets that are 1½"–3" long; each segment has 2–3 lobes. Yellow, cup-shaped flowers are ¾"–1¼" wide and have 5 petals and numerous stamens. Fruit is a seed.

Bloom season: Late winter to early summer

Range/habitat: Central and eastern North America in moist wetlands, thickets, seeps, and along stream banks

Comments: *Ranunculus* is derived from the Latin *rana* ("frog"), for the aquatic or moist habitat preference of the genus. *Hispidus* ("bristly") refers to hairs on the plant. Although the foliage is toxic to wildlife, animals such as wild turkeys, ruffed grouse, and eastern chipmunks eat the seeds.

LARGE-LEAVED AVENS

Geum macrophyllum
Rose family (Rosaceae)

Description: Perennial, with hairy stems up to 3' tall. Basal leaves with rough hairs are irregularly divided into small leaflets with a larger heart- to kidney-shaped leaflet at the top. Stem leaves have 3 deep lobes or divisions. The ½"-wide, saucer-shaped yellow flowers have 5 rounded petals surrounding a center of numerous stamens. Fruit is a seed with hooked bristles.

Bloom season: Late spring to midsummer

Range/habitat: Western and northern North America in open forests and along stream banks and forest edges

Comments: *Geum* is the Latin name for the plant. *Macrophyllum* ("large-leaved") refers to the size of the basal leaves. Northwest tribes used the plant medicinally for eyewashes, stomach ailments, and childbirth. The seed's hooked bristles catch on the fur of passing animals to aid in dispersal.

GORDON'S IVESIA

Ivesia gordonii
Rose family (Rosaceae)

Description: Perennial. Basal leaves are fernlike, with highly dissected overlapping lobes. Stems have white or sticky hairs. Flower stems bear somewhat rounded flower clusters with 10–20, ½"-long, starlike yellow flowers. Flower heads may be up to 8" wide and have 5 narrow petals and 5 stamens. Fruit is a seed.

Bloom season: Midsummer to early fall

Range/habitat: Western United States in rocky ridges, talus slopes, and mountain meadows

Comments: *Ivesia* honors Dr. Eli Ives (1779–1861), an American botanist and physician. *Gordonii* is for Alexander Gordon (1813–c. 1873) a Scottish horticulturalist and nurseryman who traveled the Oregon Trail in the late 1840s and collected the first specimen of this plant for science in 1844 along the Platte River.

PACIFIC SILVERWEED

Potentilla anserina

Rose family (Rosaceae)

Description: Perennial, with sprawling stems and reddish runners. Compound leaves are divided into 9–31 leaflets that are toothed along the margins and have white hairs below. Leaves are 3"–8" long. Bright yellow flowers, ¾"–1½" wide, are saucer-shaped and borne singly. Fruit is a flattened, oval seed.

Bloom season: Late spring through summer

Range/habitat: Western North America in coastal beaches, headlands, swales, and open areas

Comments: *Potentilla* is from the Latin *potens* ("powerful"), for the medicinal uses of the plant. *Anserina* ("related to geese") refers to the plant growing in the same habitat where geese forage. Coastal tribes ate the roots. The common name refers to the silvery color of the leaves. Some taxonomists use the synonym *Argentina anserina* for this species.

STEVE R. TURNER

ROUGH CINQUEFOIL

Potentilla norvegica

Rose family (Rosaceae)

Description: Annual, biennial, or perennial, with flowering stalks 1'–2' tall. Basal rosette of trifoliate leaves with hairy stems is about 6" across. Oval to oblong leaflets are about 2" long, coarsely toothed along the margins, and hairy on the upper surface. Upper leaves are similar but smaller. Yellow flowers arise from leaf axils or the terminal ends of the stems in loose, open clusters. The ½"-wide flowers have 5 petals and 5 triangular sepals with numerous stamens; 5 green bracts subtend the flowers. Sepals are longer than the petals. Fruit is a flattened, kidney-shaped seed.

Bloom season: Early summer to early fall

Range/habitat: Widespread across much of North America in fields, disturbed sites, and along roadsides

Comments: *Potentilla* is from the Latin *potens* ("powerful"), for the medicinal uses of the plant. *Norvegica* ("from Norway") refers to the concept that the plant may be native to Norway and spread to the New World, although modern taxonomists believe the plants to be native to the United States. Small bees and flies are common pollinators for the flowers.

ELEANOR DIETRICH

YELLOW PITCHER-PLANT

Sarracenia flava

Pitcher Plant family (Sarraceniaceae)

Description: Carnivorous perennial; leaves are 2'–3' tall. Rolled leaves resemble upright trumpets, and the upper portion flares outward to form a horizontal lid. Downward-pointing hairs and reddish nectar lines guide insects to the tube's entrance. Yellow, 5-petaled fragrant flowers are borne on a long leafless stalk, which may be 20"–36" long. The flower's strap-like petals hang over an umbrella-shaped style; stigmas are located at the tips of the "spokes." Fruit is a capsule.

Bloom season: Spring

Range/habitat: Southeastern United States in sandy bogs, marshes, savannas, and seeps

Comments: *Sarracenia* honors Dr. Michael Sarrazin (1659–1734), who reportedly sent the first specimens of this plant back to Europe for cultivation around 1700. *Flava* ("yellow") refers to the color of the flowers. The digestive juices in the leaves dissolve trapped insects and provide the plants with nutrients, especially nitrogen. Glandular hairs inside the leaves secrete nectar that also contains coniine, a toxic alkaloid that has a somewhat paralyzing effect on insects; this, along with the waxy interior surface, affects the doomed insect's mobility. In summer the plants produce long, flat "winter leaves" called phyllodia, which replace the trumpet-like tubes.

STEVE R. TURNER

CLAMMY GROUND-CHERRY

Physalis heterophylla

Nightshade family (Solanaceae)

Description: Perennial, up to 20" tall. Stems and leaves have glandular hairs that make the leaves feel "clammy." Leaves are alternate, egg-shaped with heart-shaped bases, hairy, and 2⅓"–5" long. Yellow funnel-shaped flowers are 1" wide, with 5 pale yellow petals that have purplish-brown spotting inside the throat and 5 sepals, which inflate and enclose the fruit at maturity. Fruit is a yellow, ½"-wide, round berry.

Bloom season: Early summer to early fall

Range/habitat: Primarily eastern North America, but has expanded to nearly all the US states in prairies, sandy or gravelly sites, and disturbed areas

Comments: *Physalis* is from the Greek *physalis* ("bladder or bubble"), which refers to the inflated nature of the calyx. *Heterophylla* ("different-shaped leaves") refers to the irregular shape and sizes of the leaves. The mature yellow fruits are edible, but the developing fruits contain toxic amounts of solanum. The foliage is also somewhat toxic to livestock.

BORREGOWILDFLOWERS.ORG

BUFFALO-BUR

Solanum rostratum
Nightshade family (Solanaceae)

Description: Annual, 3'–5' tall, with abundant yellow spines on the stems and leaves. Pinnately compound leaves, 1"–3½" long, are once or twice pinnate, hairy, and with wavy margins. Yellow flowers are pentagonal looking, ¾"–1½" wide, with 5 ovate-triangular petals. Anthers are different (heteranthery) and vary in size and coloration. Spiny, lance-shaped sepals continue to grow after maturity and encase the fruit. Fruit is a berrylike structure that distributes its seeds when dry.

Bloom season: Early to late summer

Range/habitat: Widespread across the United States and into central Mexico in fields, roadsides, disturbed areas, and fallow pastures

Comments: *Solanum* ("quieting") refers to related plants in the genus having narcotic properties and being used as such medicinally. *Rostratum* ("a little beak") refers to the spines. Bumblebees are common pollinators in the plant's native distribution, although the plant has expanded its range by colonizing disturbed habitats. This plant was the historical host to the Colorado potato beetle (*Leptinotarsa decemlineata*), but that beetle now prefers the domestic potato. Watermelon nightshade (*S. citrullifolium*) has bluish flowers and leaves that resemble those of a watermelon. At maturity, the plant breaks off at the stem and becomes a tumbleweed, dispersing seeds as it rolls across the landscape.

BORREGOWILDFLOWERS.ORG

COMMON GOLD-STARS

Bloomeria crocea
Brodiaea family (Themidaceae)

Description: Perennial; the corm (swollen underground stem) has a fibrous covering. Stems are up to 2' tall. A single leaf arises from the root. Starlike flowers are borne in loose clusters on long, thin stalks that resemble umbrella spokes radiating outward from a central point. The 6-petaled golden flowers have a 3-lobe style; tepals fuse together at the base to form a cuplike structure that holds nectar for pollinators. Stamens produce blue pollen. Fruit is a capsule with black seeds.

Bloom season: Mid-spring to early summer

Range/habitat: Southern California and northern Baja California in chaparral, grasslands, dry hillsides, coastal sage scrub, and open woodlands

Comments: *Bloomeria* honors Hiram Green Bloomer (1819–1874), a California botanist and one of the founders of the California Academy of Sciences. *Crocea* is from the Latin *crocum* ("saffron") and refers to the yellow color of the flowers. These plants are called "geophytes," referring to the starchy corm being a "storage organ" that is replaced each year by a new corm. The corm enables the plant to survive wildfires.

NATALIE TOLER

PITTED STRIPESEED

Piriqueta cistoides
Passionflower family (Turneraceae)

Description: Annual, 8"–12" tall, with hairy, slender stems. Elliptical to lance-shaped leaves are alternate, 1"–2" long, hairy, and with wavy margins. Bright yellow disk-shaped flowers have 5 petals, each about ½"–¾" long. Fruit is a capsule, and the seeds have longitudinal striped depressions.

Bloom season: Midsummer to early fall

Range/habitat: Southeastern United States in sandhills and open woods

Comments: *Piriqueta* is the common name for this genus in Guiana. *Cistoides* ("flowers like *Cistus*") refers to the flowers resembling those in the *Cistus,* or Rockrose, genus. The seeds are dispersed by being ejected from the capsule or by ants carrying the seeds to their nests. The flowers have developed a "heterostyly" condition, meaning some flowers have long stamens and short styles while other flowers have short stamens and long styles, all to promote cross-pollination. The common name refers to the longitudinal depressions on the long seeds.

KAREN E. ORSO

STREAM VIOLET

Viola glabella
Violet family (Violaceae)

Description: Perennial, 4"–12" tall. Heart- or kidney-shaped leaves are bright green, pointed at the tip, and toothed along the edges; basal leaves have stems 3"–8" long. A single flower is borne on a stem. Yellow flowers are ½" wide. Lower and lateral petals have purple veins; the 2 lateral petals have clumps of white hairs on their inner portions. Fruit is a capsule with brown seeds.

Bloom season: Mid-spring to midsummer

Range/habitat: Northwestern North America in forest edges, woodlands, and along streams

Comments: *Viola* ("violet-colored") is for the flower color of certain species in this genus. *Glabella* ("somewhat glabrous") refers to the leaves being mostly smooth and without hairs. The young leaves and flower buds are edible, but older flowers may cause diarrhea. Also known as the pioneer violet for its pioneering habit of growing in disturbed places.

MARGARET MARTIN

NUTTALL'S VIOLET

Viola nuttallii

Violet family (Violaceae)

Description: Perennial, about 4" tall. Lance-shaped leaves are up to 2½" long, generally 3 times longer than wide, and have short hairs along the edge. Leaves appear basal, as the flower stems arise from near the base of the plant. Yellow flowers are ⅓"–½" wide and have purple nectar lines on the lower petals; lateral petals may have a few hairs or be hairless. Fruit is a capsule.

Bloom season: Mid-spring to early summer

Range/habitat: Central United States and southern Canada in prairies, grasslands, woodlands, and rocky bluffs

Comments: *Viola* ("violet-colored") is for the flower color of certain species in this genus. *Nuttallii* honors Thomas Nuttall (1786–1859), an English botanist, zoologist, and curator of the Harvard Botanic Gardens who collected plants and wildlife in the western United States in the 1800s. The leaves and flowers are edible. This violet is a larval host for the coronis fritillary butterfly (*Speyeria coronis*). Downy yellow violet (*V. pubescens*) also has yellow flowers.

AARON GUNNAR

SLENDER YELLOW-EYED GRASS

Xyris torta

Yellow-eyed Grass family (Xyridaceae)

Description: Perennial, up to 40" tall, with twisted stems. Basal, grasslike leaves are 8"–20" long, hairless, with entire margins, and have red or purplish bases. A cylindrical flower head, ⅜"–1" long, is borne at the tip of the flowering stalk. The head is subtended by several brown, scalelike bracts that spiral around the base. Individual yellow flowers, about ⅓" wide with 3 ragged petals, arise from the axils of the bracts. There are 3 fertile and 3 infertile stamens. Fruit is a capsule with translucent seeds.

Bloom season: Midsummer to early fall

Range/habitat: Central and eastern United States in fens, bogs, shores, wetlands, and peatlands

Comments: *Xyris* ("iris") refers to the flowers resembling those of the Iris family. *Torta* ("twisted") refers to the twisted flowering stems; another common name is twisted yellow-eyed grass. The plants arise from a chestnut-brown bulb.

GLOSSARY

Alternate—Placed singly along a stem or axis, one after another; often used in reference to leaf arrangement on a stem (*see* Opposite).
Annual—A plant completing its life cycle, from seed germination to production of new seeds, within a year.
Axil—The area created on the upper side of the angle between a leaf and stem.
Basal—At the base or bottom of; generally used in reference to leaves arranged at the base of a plant.
Biennial—A plant completing its life cycle in two years and normally not producing flowers during the first year.
Bract—Reduced or modified leaf, often associated with flowers.
Bristle—A stiff hair, usually erect or curving away from its attachment point.
Bulb—Underground plant part derived from a short, usually rounded, shoot that is covered with fleshy scales.
Calyx—The outer set of flower parts, composed of the sepals, which may be separate or joined together; usually green.
Capsule—A dry fruit that releases seeds through splits or holes.
Compound leaf—A leaf that is divided into two to many leaflets, each of which may resemble a completed leaf but which lacks buds (*see* illustration page 5).
Corolla—The set of flower parts interior to the calyx and surrounding the stamens, composed of the petals, which may be free or united; often brightly colored.
Deciduous—Referring to broad-leaved trees or shrubs that drop their leaves at the end of each growing season, as contrasted with plants that retain their leaves throughout the year (*see* Evergreen).
Disk flowers—Small, tubular flowers in the central portion of the flower head of many plants in the Aster family (Asteraceae) (*see* illustrations page 8).
Elliptical (leaf shape)—*See* illustration page 6.
Entire (leaf margin)—*See* illustration page 6.
Evergreen—Referring to plants that bear green leaves throughout the year, as contrasted with plants that lose their leaves at the end of the growing season (*see* Deciduous).

Family—A group of plants having biologically similar features, such as flower anatomy, fruit type, etc.

Flower head—As used in this guide, a dense and continuous group of flowers, without obvious branches or space between them; used especially in reference to the Aster family (Asteraceae).

Genus—A group of closely related species, such as the genus *Penstemon* encompassing the penstemons (*see* Specific epithet).

Herbaceous—Referring to any nonwoody plant; often reserved for wildflowers or forbs.

Hood—Curving or folded, petallike structure interior to the petals and exterior to the stamen in the Milkweed family (Apocynaceae). Since most milkweeds have reflexed petals, the hoods are typically the most prominent feature of the flowers.

Inflorescence—Generally a cluster of flowers, although there are many terms to specifically describe the arrangement of flowers on a plant.

Involucre—A distinct series of bracts or leaves that subtend a flower or cluster of flowers. Often used in the description of Aster family (Asteraceae) flower heads.

Keel—A sharp lengthwise fold or ridge, referring particularly to the two fused petals forming the lower lip in many flowers of the Pea family (Fabaceae).

Lance (leaf shape)—*See* illustration page 6.

Leaflet—A distinct, leaflike segment of a compound leaf.

Linear (leap shape)—*See* illustration page 6.

Lobe—A segment of an incompletely divided plant part, typically rounded; often used in reference to the leaves.

Midrib—The center or main vein of a leaf.

Node—The region of the stem where one or more leaves are attached. Buds are commonly borne at the node in the axils of leaves.

Nutlet—A descriptive term for small nutlike fruits. Used to describe the separate lobes of a mature ovary in the Borage (Boraginaceae) and Mint (Lamiaceae) families.

Oblong (leaf shape)—*See* illustration page 6.

Opposite—Paired directly across from one another along a stem or axis (*see* Alternate).

Ovary—The portion of the flower where the seeds develop; usually a swollen area below the style and stigma.

Pappus—In the Aster family (Asteraceae), the modified limb of the calyx is the pappus.

Parallel—Side by side, approximately the same distance apart for the entire length; often used in reference to veins or edges of leaves.

Perennial—A plant that normally lives for three or more years.

Petal—Component part of the corolla; often the most brightly colored and visible part of the flower.

Petiole—The stalk of a leaf. The length of the petiole may be used in leaf descriptions.

Pinnate—Referring to a compound leaf, like many members of the Pea family (Fabaceae), where smaller leaflets are arranged along either side of a common axis.

Pistil—The seed-producing, or female, part of a flower, consisting of the ovary, style (if present), and stigma.

Pollen—Tiny, often powdery male reproductive cells formed in the stamens and typically necessary for seed production.

Ray flower—Flower in the Aster family (Asteraceae) with a single, strap-shaped corolla, resembling one flower petal. Several to many ray flowers may surround the disk flowers in a flower head, or in some species, such as dandelions, the flower heads may be composed entirely of ray flowers (*see* illustration page 8).

Rosette—A dense cluster of basal leaves from a common underground part, often in a flattened, circular arrangement.

Scale—Any thin, membranous body that somewhat resembles the scales of fish or reptiles.

Sepal—Component part of the calyx; typically green, but sometimes enlarged and brightly colored.

Shrub—A perennial woody plant of relatively low height; typically with several stems arising from or near the ground.

Silique—The podlike fruiting body of plants in the Mustard family (Brassicaceae) that is longer than wide.

Simple leaf—A leaf that has a single leaflike blade, although this may be lobed or divided.

Spatula (leaf shape)—*See* illustration page 6.

Specific epithet—The second portion of a scientific name, identifying a particular species; for instance, in **Columbia lily** (*Lilium columbianum*) the specific epithet is "*columbianum*."

Spike—An elongate, unbranched cluster of stalkless or nearly stalkless flowers.

Stalk—As used here, the stem supporting the leaf, flower, or flower cluster.

Stalkless—Lacking a stalk; a stalkless leaf is attached directly to the stem at the leaf base.

Stamen—The main unit of a flower, which produces the pollen; typically consisting of a long filament with a pollen-producing tip.

Standard—The usually erect, spreading upper petal in many flowers of the Pea family (Fabaceae).

Stigma—Portion of the pistil between the ovary and the stigma; typically a slender stalk.

Style—The stalklike part of the pistil that connects the ovary to the stigma.

Subtend—To be situated below or beneath, often encasing or enclosing something.

Toothed—Bearing teeth, or sharply angled projections, along the edge.

Variety—A group of plants within a species with a distinct range, habitat, or structure.

Whorl—Three or more parts attached at the same point along a stem or axis and often surrounding the stem.

Wings—The two side petals flanking the keel in many flowers of the Pea family (Fabaceae).

RESOURCES

Viewing wildflowers in their natural habitats is highly encouraged. There are numerous private, city, state, and federal parks, forests, and preserves to observe flowering plants and pollinator activity. FalconGuides offers numerous wildflower field guides to specific areas. Numerous conservation organizations and Native Plant Society chapters provide information or field trips to view wildflowers as well.

An internet search will also provide a wealth of information regarding times and places to enjoy wildflowers. Some sites provide "Wildflower Hotlines" similar to those used by birders. Here are just a few of the many online resources available:

NATIONAL PARK SERVICE
nps.gov

WILDFLOWER SEARCH
wildflowersearch.org

NATIONAL FOREST SERVICE
www.fs.fed.us/wildflowers/viewing/index.php

NATIVE PLANT SOCIETY
Varies (search by state or local chapter)

USDA NATURAL RESOURCE CONSERVATION SERVICE
plants.usda.gov

THE NATURE CONSERVANCY
tnc.org

LADY BIRD JOHNSON WILDFLOWER CENTER
wildflower.org

WILDFLOWER IDENTIFICATION WEBSITES
identifythatplant.com

WILDFLOWERS OF THE UNITED STATES
uswildflowers.com

DESERTUSA WILDFLOWER REPORTS
desertusa.com

ONTARIO WILDFLOWERS
ontariowildflowers.com

THEODORE PAYNE WILDFLOWER HOTLINE (Southern California)
theodorepayne.org/learn/wildflower-hotline/

MINNESOTA WILDFLOWERS
minnesotawildflowers.info

SOUTHWEST COLORADO WILDFLOWERS
swcoloradowildflowers.com

OREGON FLORA PROJECT
oregonflora.org

ANZA-BORREGO DESERT WILDFLOWERS
borregowildflowers.org

INDEX

ABOUT THE AUTHOR

Damian Fagan obtained his BS in botany from the University of Washington. He is a former National Park Service ranger, The Nature Conservancy preserve manager, field biologist, and communications manager for the High Desert Museum. He currently writes about the natural world for a variety of publications. Damian lives in the Pacific Northwest and is best identified as the guy with binoculars and a camera slung around his neck.